CHOOSING AND USING STATISTICS

Learning F

Choosing and Using Statistics:
A Biologist's Guide

Calvin Dytham

Department of Biology, University of York

Third Edition

WILEY-BLACKWELL

A John Wiley & Sons, Ltd., Publication

This edition first published 2011, © 1999, 2003 by Blackwell Science, 2011 by Calvin Dytham

Blackwell Publishing was acquired by John Wiley & Sons in February 2007. Blackwell's publishing program has been merged with Wiley's global Scientific, Technical and Medical business to form Wiley-Blackwell.

Registered Office:
John Wiley & Sons Ltd, The Atrium, Southern Gate, Chichester, West Sussex, PO19 8SQ, UK

Editorial Offices:
9600 Garsington Road, Oxford, OX4 2DQ, UK
The Atrium, Southern Gate, Chichester, West Sussex, PO19 8SQ, UK
111 River Street, Hoboken, NJ 07030-5774, USA

For details of our global editorial offices, for customer services and for information about how to apply for permission to reuse the copyright material in this book please see our website at www.wiley.com/wiley-blackwell.

The right of the author to be identified as the author of this work has been asserted in accordance with the UK Copyright, Designs and Patents Act 1988.

Library of Congress Cataloging-in-Publication Data

Dytham, Calvin.
    Choosing and using statistics : a biologist's guide / by Calvin Dytham. – 3rd ed.
       p. cm.
    Includes bibliographical references and index.
    ISBN 978-1-4051-9838-7 (hardback) – ISBN 978-1-4051-9839-4 (pbk.)
1. Biometry.   I. Title.
    QH323.5.D98 2011
    001.4'22–dc22
                                                2010030975

A catalogue record for this book is available from the British Library.

This book is published in the following electronic format: ePDF 978-1-4443-2843-1

Set in 9.5/12pt Berling by SPi Publisher Services, Pondicherry, India
Printed and bound in Malaysia by Vivar Printing Sdn Bhd

5   2014

Contents

7 The tests 1: tests to look at differences 72

Preface

My aim was to produce a statistics book with two characteristics: to assume that the reader is using a computer to analyse data and to contain absolutely no equations.

This is a handbook for biologists who want to process their data through a statistical package on the computer, to select the most appropriate methods and extract the important information from the, often confusing, output that is produced. It is aimed, primarily, at undergraduates and masters students in the biological sciences who have to use statistics in practical classes and projects. Such users of statistics don't have to understand exactly how the test works or how to do the actual calculations. These things are not covered in this book as there are more than enough books providing such information already. What is important is that the right statistical test is used and the right inferences made from the output of the test. An extensive key to statistical tests is included for the former and the bulk of the book is made up of descriptions of how to carry out the tests to address the latter.

In several years of teaching statistics to biology students it is clear to me that most students don't really care how or why the test works. They do care a great deal that they are using an appropriate test and interpreting the results properly. I think that this is a fair aim to have for occasional users of statistics. Of course, anyone going on to use statistics frequently should become familiar with the way that calculations manipulate the data to produce the output as this will give a better understanding of the test.

If this book has a message it is this: think about the statistics *before* you collect the data! So many times I have seen rather distraught students unable to analyse their precious data because the experimental design they used was inappropriate. On such occasions I try to find a compromise test that will make the best of a bad job but this often leads to a weaker conclusion than might have been possible if more forethought had been applied from the outset. There is no doubt that if experiments or sampling strategies are designed with the statistics in mind better science will result.

Statistics are often seen by students as the 'thing you must do to data at the end'. Please try to avoid falling into this trap yourself. Thought experiments producing dummy data are a good way to try out experimental designs and are much less labour-intensive than real ones!

Although there are almost no equations in this book I'm afraid there was no way to totally avoid statistical jargon. To ease the pain somewhat, an extensive Glossary and key to symbols are included. So when you are navigating your way through the key to choosing a test you should look up any words you don't understand.

In this book I have given extensive instructions for the use of four commonly encountered software packages: SPSS, R, Excel and MINITAB. However, the key to choosing a statistical test is not at all package-specific, so if you use a software package other than the four I focus on or if you are using a calculator you will still be able to get a good deal out of this book.

If every sample gave the same result there would be no need for statistics. However, all aspects of biology are filled with variation. It is statistics that can be used to penetrate the haze of experimental error and the inherent variability of the natural world to reach the underlying causes and processes at work. So, try not to hate statistics, they are merely a tool that, when used wisely and properly, can make the life of a biologist much simpler and give conclusions a sound basis.

The third edition

In the 8 years since I wrote the second edition of this book there have, of course, been several new versions of the software produced. I have received many comments about the previous editions and I am grateful for the many suggestions on how to improve the text and coverage. Requests to add further statistical packages have been the most common suggestion for change. There was surprisingly little consensus on the packages to add for the second edition, but since 2000 the freely available, and very powerful, package R has become extremely widely used so I have added that to the mix this time.

How to use this book

This is definitely not a book that should be read from cover to cover. It is a book to refer to when you need assistance with statistical analysis, either when choosing an appropriate test or when carrying it out. The basics of statistical analysis and experimental design are covered briefly but those sections are intended mostly as a revision aid, or to outline of some of the more important concepts. The reviews of other statistics books may help you choose those that are most appropriate for you if you want or need more details.

The heart of the book is the key. The rest of the book hinges on the key, explaining how to carry out the tests, giving assistance with the statistical terms in the Glossary or giving tips on the use of computers and packages.

Packages used

MINITAB® version 15, MINITAB Inc.
SPSS® versions 16 and 17, SPSS Inc.
Excel™ version 2007 and 2008 for Mac, Microsoft Corporation
Running on:
Windows® versions XP, 2000, 7 and Vista, Microsoft Corporation
Mac OS 10, Apple Inc.

Example data

In the spirit of dummy data collection, all example data used throughout this book have been fabricated. Any similarity to data alive or dead is purely coincidental.

Acknowledgements for the first edition

Thanks to Sheena McNamee for support during the writing process, to Andrea Gillmeister and two anonymous reviewers for commenting on an early version of the manuscript and to Terry Crawford, Jo Dunn, David Murrell and Josephine Pithon for recommending and lending various books. Thanks also to Ian Sherman and Susan Sternberg at Blackwell and to many of my colleagues who told me that the general idea of a book like this was a sound one. Finally, I would especially like to thank the students at the University of York, UK, who brought me the problems that provided the inspiration for this book.

Acknowledgements for the second edition

Thanks to all the many people who contacted me with suggestions and comments about the first edition. I hope you can see that many of the corrections and improvements have come directly from you. Five anonymous reviewers provided many useful comments about the proposal for a second edition. Thanks to Sarah Shannon, Cee Brandston, Katrina McCallum and many others at Blackwell for seeing this book through and especially for producing a second superb and striking cover. S'Albufera Natural Parc and Nick Riddiford provided a very convenient bolt-hole for writing. Once again, I give special thanks to Sheena and to my colleagues, PhD students and undergraduate students at the University of York. Finally, thanks to everyone on the MRes EEM course over the last 4 years.

Acknowledgements for the third edition

It's been thanks to the pushing of Ward Cooper at Wiley-Blackwell and Sheena McNamee that this third edition has seen the light of day. Thanks to Emma Rand, Olivier Missa and Frank Schurr for encouraging me to enter the brave new world of R. Thanks to Nik Prowse for guiding me through the final editing.

Calvin Dytham,
York 1998, 2002 and 2010

Eight steps to successful data analysis

1

This is a very simple sequence that, if you follow it, will integrate the statistics you use into the process of scientific investigation. As I make clear here, statistical tests should be considered *very early* in the process and not left until the end.

1 Decide what you are interested in.

2 Formulate a hypothesis or several hypotheses (see Chapters 2 and 3 for guidance).

3 Design the experiment, manipulation or sampling routine that will allow you to test the hypotheses (see Chapters 2 and 4 for some hints on how to go about this).

4 *Collect dummy data* (i.e. make up approximate values based on what you expect to obtain). The collection of 'dummy data' may seem strange but it will convert the proposed experimental design or sampling routine into something more tangible. The process can often expose flaws or weaknesses in the data-collection routine that will save a huge amount of time and effort.

5 Use the key presented in Chapter 3 to guide you towards the appropriate test or tests.

6 Carry out the test(s) using the dummy data. (Chapters 6–9 will show you how to input the data, use the statistical packages and interpret the output.)

7 If there are problems go back to step 3 (or 2); otherwise, proceed to the collection of real data.

8 Carry out the test(s) using the real data. Report the findings and/or return to step 2.

I implore you to use this sequence. I have seen countless students who have spent a long time and a lot of effort collecting data only to find that the experimental or sampling design was not quite right. The test they are forced to use is much less powerful than one they could have used with only a slight change in the experimental design. This sort experience tends to turn people away from statistics and become 'scared' of them. This is a great shame as statistics are a hugely useful and vital tool in science.

The rest of the book follows this eight-step process but you should use it for guidance and advice when you become unsure of what to do.

Choosing and Using Statistics: A Biologist's Guide, 3rd Edition. By Calvin Dytham.
Published 2011 by Blackwell Publishing Ltd.

2

The basics

The aim of this chapter is to introduce, in rather broad terms, some of the recurring concepts of data collection and analysis. Everything introduced here is covered at greater length in later chapters and certainly in the many statistics textbooks that aim to introduce statistical theory and experimental design to scientists.

The key to statistical tests in the next chapter assumes that you are familiar with most of the basic concepts introduced here.

Observations

These are the raw material of statistics and can include anything recorded as part of an investigation. They can be on any scale from a simple 'raining or not raining' dichotomy to a very sophisticated and precise analysis of nutrient concentrations. The type of observations recorded will have a great bearing on the type of statistical tests that are appropriate.

Observations can be simply divided into three types: *categorical* where the observations can be in a limited number of categories which have no obvious scale (e.g. 'oak', 'ash', 'elm'); *discrete* where there is a real scale but not all values are possible (e.g. 'number of eggs in a nest' or 'number of species in a sample') and *continuous* where any value is theoretically possible, only restricted by the measuring device (e.g. lengths, concentrations).

Different types of observations are considered in more detail in Chapter 5.

Hypothesis testing

The cornerstone of scientific analysis is hypothesis testing. The concept is rather simple: almost every time a statistical test is carried out it is testing the probability that a hypothesis is correct. If the probability is small then the hypothesis is deemed to be untrue and it is rejected in favour of an alternative. This is done in what seems to be a rather upside down way as the test is always of what is

Choosing and Using Statistics: A Biologist's Guide, 3rd Edition. By Calvin Dytham.
Published 2011 by Blackwell Publishing Ltd.

called the null hypothesis rather than the more interesting hypothesis. The null hypothesis is the hypothesis that nothing is going on (it is often labelled as H_0). For example, if the weights of bulbs for two cultivars of daffodils were being investigated, the null hypothesis would be that there is no weight difference between cultivars: 'the weights of the two groups of bulbs are the same' or, more correctly, 'the two groups of bulbs are samples from a larger population with the same distribution'. A statistical test is carried out to find out how likely that null hypothesis is to be true. If we decide to reject the null hypothesis we must accept the alternative, more interesting, hypothesis (H_1) that: 'the weights of bulbs for the two cultivars are different' or, more correctly, that 'the groups are samples from populations with different distributions'.

P-values

The *P*-value is the bottom line of most statistical tests. (Incidentally, you may come across it written in upper or lower case, italic or not: e.g. P value, *P*-value, *p* value or p-value.) It is the probability of seeing data this extreme or more extreme if the null hypothesis is true. So if a *P*-value is given as 0.06 it indicates that you have a 6% chance of seeing data like this if the null hypothesis is true. In biology it is usual to take a value of 0.05 or 5% as the critical level for the rejection of a hypothesis. This means that providing a hypothesis has a less than one in 20 chance of being true we reject it. As it is the null hypothesis that is nearly always being tested we are always looking for low *P*-values to reject this hypothesis and accept the more interesting alternative hypothesis.

Clearly the smaller the *P*-value the more confident we can be in the conclusions drawn from it. A *P*-value of 0.0001 indicates that if the null hypothesis is true the chance of seeing data as extreme or more extreme than that being tested is one in 10 000. This is much more convincing than a marginal $P = 0.049$.

P-values and the types of errors that are implicitly accepted by their use are considered further in Chapter 4.

Sampling

Observations have to be collected in some way. This process of data acquisition is called sampling. Although there are almost as many different methods that can be used for sampling as there are possible things to sample, there are some general rules. One of the most obvious is that a large number of observations is usually better than a small number. Balanced sampling is also important (i.e. when comparing two groups take the same number of observations from each group).

Most statistical tests assume that samples are taken at random. This sounds easy but is actually quite difficult to achieve. For example, if you are sampling beetles from pit-fall traps the sample may seem totally random but in fact is quite biased towards those species that move around the most and fail to avoid the traps. Another common bias is to chose a point at random and then measure the nearest individual to that point, assuming that this will produce a random sample. It will not be random at all as isolated individuals and those at the edges of clumps are more likely to be selected than those in the middle. There are methods available to reduce problems associated with non-random sampling but the first step is to be aware of the problem.

A further assumption of sampling is that individuals are either only measured once or they are all sampled on several occasions. This assumption is often violated if, for example, the same site is visited on two occasions and the same individuals or clones are inadvertently remeasured.

The sets of observations collected are called variables. A variable can be almost anything it is possible to record as long as different individuals can be assigned different values.

Some of the problems of sampling are considered in Chapter 4.

Experiments

In biology many investigations use experiments of some sort. An experiment occurs when anything is altered or controlled by the investigator. For example, an investigation into the effect of fertilizer on plant growth will use a control plot (or several control plots) where there is no fertilizer added and then one or more plots where fertilizer has been added at known concentrations set by the investigators. In this way the effect of fertilizer can be determined by comparison of the different concentrations of fertilizer. The condition being controlled (e.g. fertilizer) is usually called a factor and the different levels used called treatments or factor levels (e.g. concentrations of fertilizer). The design of this experiment will be determined by the hypothesis or hypotheses being investigated. If the effect of the fertilizer on a particular plant is of interest then perhaps a range of different soil types might be used with and without fertilizer. If the effect on plants in general is of interest then an experiment using a variety of plants is required, either in isolation or together. If the optimum fertilizer treatment is required then a range of concentrations will be applied and a cost-benefit analysis carried out.

More details and strategies for experimental design are considered in Chapter 4.

Statistics

In general, statistics are the results of manipulation of observations to produce a single, or small number of results. There are various categories of statistics depending on the type of summary required. Here I divide statistics into four categories.

Descriptive statistics

The simplest statistics are summaries of data sets. Simple summary statistics are easy to understand but should not be overlooked. These are not usually considered to be statistics but are in fact extremely useful for data investigation. The most widely used are measures of the 'location' of a set of numbers such as the mean or median. Then there are measures of the 'spread' of the data, such as the standard deviation. Choice of appropriate descriptive statistic and the best way of displaying the results are considered in Chapters 5 and 6.

Tests of difference

A familiar question in any field of investigation is going to be something like 'is this group different from that group?'. A question of this kind can then be turned into a null hypothesis with a form: 'this group and that group are not different'. To answer this question, and test the null hypothesis, a statistical test of difference is required. There are many tests that all seem to answer the same type of question but each is appropriate when certain types of data are being considered. After the simple comparison of two groups there are extensions to comparisons of more than two groups and then to tests involving more than one way of dividing the individuals into groups. For example, individuals could be assigned to two groups by sex and also into groups depending on whether they had been given a drug or not. This could be considered as four groups or as what is known as a factorial test, where there are two factors, 'sex' and 'drug', with all combinations of the levels of the two factors being measured in some way. Factorial designs can become very complicated but they are very powerful and can expose subtleties in the way the factors interact that can never be found though investigation of the data using one factor at a time.

Tests of difference can also be used to compare variables with known distributions. These can be statistical distributions or derived from theory. Chapter 7 considers tests of difference in detail.

Tests of relationships

Another familiar question that arises in scientific investigation is in the form 'is A associated with B?'. For example, 'is fat intake related to blood pressure?'. This type of question should then be turned into a null hypothesis that 'A is not associated with B' and then tested using one of a variety of statistical tests. As with tests of difference there are a many tests that seem to address the same type of problem, but again each is appropriate for different types of data.

Test of relationships fall into two groups, called correlation and regression, depending on the type of hypothesis being investigated. Correlation is a test to measure the degree to which one set of data varies with another: it *does not* imply that there is any cause-and-effect relationship. Regression is used to fit a relationship between two variables such that one can be predicted from the other. This *does* imply a cause-and-effect relationship or at least an implication that one of the variables is a 'response' in some way. So in the investigation of fat intake and blood pressure a strong positive correlation between the two shows an association but does not show cause and effect. If a regression is used and there is a significant positive regression line, this would imply that blood pressure can be predicted using fat intake *or*, if the regression uses the fat intake as the 'response', that fat intake can be predicted from blood pressure.

There are many additional techniques that can be employed to consider the relationships between more than two sets of data. Tests of relationships are described in Chapter 8.

Tests for data investigation

A whole range of tests is available to help investigators explore large data sets. Unlike the tests considered above, data investigation need not have a hypothesis for testing. For example, in a study of the morphology of fish there may be many fin measures from a range of species and sites that offer far too many potential hypotheses for investigation. In this case the application of a multivariate technique may show up relationships between individuals, help assign unknown specimens to categories or just suggest which hypotheses are worth further consideration.

A few of the many different techniques available are considered in Chapter 9.

Choosing a test: a key

3

I hope that you are not reading this chapter with your data already collected and the experiment or sampling programme 'finished'. If you have finished collecting your data I strongly advise you to approach your next experiment or survey in a different way. As you will see below, I hope that you will be using this key *before* you start collecting real data.

Remember: eight steps to successful data analysis

1 Decide what you are interested in.
2 Formulate a hypothesis or hypotheses.
3 Design the experiment or sampling routine.
4 Collect dummy data. Make up approximate values based on what you expect.
5 *Use the key here to decide on the appropriate test or tests.*
6 Carry out the test(s) using the dummy data.
7 If there are problems go back to step 3 (or 2), otherwise collect the real data.
8 Carry out the test(s) using the real data.

The art of choosing a test

It may be a surprising revelation, but choosing a statistical test is not an exact science. There is nearly always scope for considerable choice and many decisions will be made based on personal judgements, experience with similar problems or just a simple hunch. There are many circumstances under which there are several ways that the data could be analysed and yet each of the possible tests could be justified.

A common tendency is to force the data from your experiment into a test you are familiar with even if it is not the best method. Look around for different tests that may be more appropriate to the hypothesis you are testing. In this way you will expand your statistical repertoire and add power to your future experiments.

Choosing and Using Statistics: A Biologist's Guide, 3rd Edition. By Calvin Dytham. Published 2011 by Blackwell Publishing Ltd.

A key to assist in your choice of statistical test

Starting at step 1 in the list above move through the key following the path that best describes your data. If you are unsure about any of the terms used then consult the glossary or the relevant sections of the next two chapters. This is not a true dichotomous key and at several points there are more than two routes or end points.

There may be several end points appropriate to your data that result from this key. For example you may wish to know the correct display method for the data and then the correct measure of dispersion to use. If this is the case, go through the key twice.

All the tests and techniques mentioned in the key are described in later chapters.

Italics indicate instructions about what you should do.

Numbers in brackets indicate that the point in the key is something of a compromise destination.

There are several points where rather arbitrary numbers are used to determine which path you should take. For example, I use 30 different observations as the arbitrary level at which to split continuous and discontinuous data. If your data set falls close to this level you should not feel constrained to take one path if you feel more comfortable with the other.

---

1. Testing a clear hypothesis and associated null hypothesis (e.g. $H_1 =$ blood glucose level is related to age and $H_0 =$ blood glucose is not related to age). 25

   Not testing any hypothesis but simply want to present, summarize or explore data. 2

2. Methods to summarize and display the data required. 3

   Data exploration for the purpose of understanding and getting a feel for the data or perhaps to help with formulation of hypotheses. For example, you may wish to find possible groups within the data (e.g. 10 morphological variables have been taken from a large number of carabid beetles; the multivariate test may establish whether they can be divided into separate taxa). 60

3. There is only one collected variable under consideration (e.g. the only variable measured is brain volume although it may have been measured from several different populations). 4

   There is more than one measured variable (e.g. you have measured the number of algae per millilitre *and* the water pH in the same sample). 24

4. The data are discrete; there are fewer than 30 different values (e.g. number of species in a sample). 5

The data are continuous; there are more than 29 different values
(e.g. bee wing length measured to the nearest 0.01 mm).
(*Note*: the distinction between the above is rather arbitrary.)   16

5 There is only one group or sample (e.g. all measurements taken   6
   from the same river on the same day).
   There is more than one group or sample (e.g. you have measured   15
   the number of antenna segments in a species of beetle and have
   divided the sample according to sex to give two groups).

6 A graphical representation of the data is required.   7
   A numerical summary or description of the data required.   11

7 A display of the whole distribution is required.   8
   Crude display of position and spread of data is required: *use a box
   and whisker display to show medians, range and inter-quartile range,
   page 49 (also known as a box plot).*

8 Values have real meaning (e.g. number of mammals caught per night).   10
   Values are arbitrary labels that have no real sequence (e.g. different   9
   vegetation-type classifications in an area of forest).

9 There are fewer than 10 different values or classifications: *draw a
   pie chart, page 52. Ensure that each segment is labelled clearly and
   that adjacent shading patterns are as distinct as possible. Avoid using
   three-dimensional or shadow effects, dark shading or colour. Do not
   add the proportion in figures to the 'piece' of the pie as this information
   is redundant.*
   There are 10 or more different values or classifications: *amalgamate
   values until there are fewer than 10 or divide the sample to produce
   two sets each with fewer than 10 values. Ten is a level above which it
   is difficult to distinguish different sections of the pie or to have
   sufficiently distinct shading patterns.*

10 There are more than 20 different values: *amalgamate values to
   produce around 12 classes (almost certainly done automatically by
   your package) and draw a histogram, page 51. Put classes on the
   x-axis, frequency of occurrence (number of times the value occurs) on
   the y-axis, with no gaps between bars. Do not use three-dimensional or
   shadow effects.*
   There are 20 or fewer different values: *draw a bar chart, page 51.
   Each value should be represented on the x-axis. If there are few classes,
   extend the range to include values not in the data set at either side,
   frequency of occurrence (number of times the value occurs) on y-axis.
   Gaps should appear between bars, unless the variable is clearly
   supposed to be continuous; do not use three-dimensional or shadow
   effects.*

11 You want a measure of position (mean is the one used most   12
    commonly).
    You want a measure of dispersion or spread (standard deviation   13
    and confidence intervals are the most commonly used).
    You want a measure of symmetry or shape of the distribution.   14
    (*Note*: you will probably want to go for at least one measure of
    position and another of spread in most cases.)

12 Variable is definitely discrete, usually restricted to integer values
    smaller than 30 (e.g. number of eggs in a clutch): *calculate the
    median, page 53.*
    Variable should be continuous but has only a few different values
    due to accuracy of measurement (e.g. bone length measured to the
    nearest centimetre): *calculate the mean, page 53.*
    If you are particularly interested in the most commonly occurring
    response: *calculate the mode, page 53, in addition to either the mean
    or median.*

13 A very rough measure of spread is required: *calculate the range, page
    55 (note that this measure is very biased by sample size and is rarely
    a useful statistic).*
    You are particularly interested in the highest and/or lowest values:
    calculate the range, page 55.
    Variable should be continuous but has only a few values due to
    accuracy of measurement: *calculate the standard deviation, page 55.*
    Variable is discrete or has an unusual distribution: *calculate the
    interquartile range, page 55.*

14 Variable should be continuous but has only a few values due to
    accuracy of measurement: *calculate the skew (g_1), page 57.*
    Observations are discrete or you have already calculated the
    interquartile range and the median: *the relative size of the
    interquartile range above and below the median provides a measure of
    the symmetry of the data.*

15 You have not established the appropriate technique for a single   (6)
    sample: *go back to 6 to find the appropriate techniques for each group.
    You should find that the same is correct for each sample or group.*
    The samples can be displayed separately: *go back to 7 and choose the   (7)
    appropriate style. So that direct comparisons can be made, be sure to
    use the same scales (both x-axis and y-axis) for each graph. Be
    warned that packages will often adjust scales for you. If this happens
    you must force the scales to be the same.*
    The samples are to be displayed together on the same graph: *use a
    chart with a box plot for each sample and the x-axis representing the
    sample number, page 62. Ensure that there is a clear space between
    each box plot.*

16 There is only a data set from one group or sample.   17
  The data have been collected from more than one group or sample   23
  (e.g. you have measured the mass of each individual of a single
  species of vole from one sample and have divided the sample
  according to sex).

17 A graphical representation of the data is required.   18
  A numerical summary or descriptive statistics are required.   19

18 A display of the whole distribution is required: *group to produce
  around 12–20 classes and draw a histogram, page 51 (probably done
  automatically by your package). Put classes on the x-axis, frequency of
  occurrence (number of times the value occurs within the class) on the
  y-axis, with no gaps between bars and no three-dimensional or shadow
  effects. Even-sized classes are much easier for a reader to interpret.
  Data with an unusual distribution (e.g. there are some extremely high
  values well away from most of the observations) may require
  transformation before the histogram is attempted.*
  A crude display of position and spread of the data is required: *the
  'error bar' type of display is unusual for a single sample but common
  for several samples. There is a symbol representing the mean and a
  vertical line representing range of either the 95% confidence interval or
  the standard deviation, page 63.*

19 You want a measure of position (mean is the most common).   20
  You want a measure of dispersion (spread).   21
  You want a measure of symmetry or shape of the distribution.   22
  You wish to determine whether the data are normally distributed:
  *carry out a Kolmogorov–Smirnov test, page 86, an Anderson–Darling
  test, page 89, a Shapiro–Wilk test, page 90, or a chi-square goodness of
  fit, page 75.*
  (*Note:* you probably require one of each of the above for a full
  summary of the data.)

20 Unless the variable is definitely discrete or is known to have an odd
  distribution (e.g. not symmetrical): *calculate the mean, page 53.*
  If the data are known to be discrete or the data set is to be
  compared with other, discrete data with fewer possible values:
  calculate the median, page 53.
  If you are particularly interested in the most commonly occurring
  value: *calculate the mode, page 53,* in addition to *the mean or median.*

21 If the data are continuous and approximately normally distributed
  and you require an estimate of the spread of data: *calculate the
  standard deviation (SD), page 55. (Note: standard deviation is the
  square root of variance and is measured in the same units as the
  original data.)*

If you have previously calculated the mean and require an estimate of the range of possible values for the mean: *calculate 95% confidence limits for the mean, page 56 (a.k.a. 95% confidence interval or 95% CI).*
A very rough measure of spread required: *calculate the range, page 55. (Note that this measure is very biased by sample size and is rarely a useful statistic in large samples.)*
If you have a special interest in the highest and or lowest values in the sample: *calculate the range, page 55.*
If the data are known to be discrete or are to be compared with other, discrete, data or if you have previously calculated the median: *calculate the interquartile range, page 55.*
(*Note:* many people use *standard error (SE)* as a measure of spread. I think that the main reason for this is that it is smaller than either SD or 95% CI rather than for any statistical reason. Do not use SE for this purpose unless you are making a comparison to previously calculated SEs.)

22 If the data are continuous and normally distributed and you require an unbiased estimate of the symmetry of the data: *calculate the skewness/asymmetry of the data (g_1), page 57. Skew is only worth calculating in samples with more than 30 observations.*
If the data are continuous and normally distributed, you have calculated skewness and you require an estimate of the 'shape' of the distribution of the data: *calculate the kurtosis (g_2), page 57. (This is rarely required as a graphical representation will give a better understanding of the shape of the data. Kurtosis is only really worth calculating in samples with more than 100 observations.)*
If you have already calculated the interquartile range and the median: *re-examine the interquartile range. The relative size of the interquartile range above and below the median provides a measure of the symmetry of the data.*

23 You have not established the appropriate technique for a single sample: *go back to 16 to find the appropriate techniques for each of the groups. You should find the same is appropriate for each sample or group.*   (16)
The samples can be displayed separately: *go back to 17 and choose the appropriate style. So that direct comparisons can be made, be sure to use the same scales (both x-axis and y-axis) for each graph. Be warned that statistical packages will often adjust scales for you.*   (17)
The samples are to be displayed together on the same graph: *use a chart with an 'error bar' (showing the mean and a measure of spread) for each sample and the x-axis representing the sample number/site. Do not join the means unless intermediate samples would be possible (i.e. don't join means from samples divided by sex or species but do join those representing temperature, if the intervals between different sample temperatures are even).*

24 If each variable is to be considered separately: *go back to 4 and*   (4)
consider each variable in turn.
Two variables only: *a two-dimensional scatterplot can be drawn,*
page 64. The choice of variable for x- *and* y-axes *is free but if you*
suspect a possibility of 'cause' and 'effect' the 'cause' should always be
on the x-axis. *Do not draw a line of best fit even if it is offered by the*
package unless the situation is appropriate and you have carried out
a regression analysis.
Three variables: *a three-dimensional scatterplot can be drawn,*
page 68. It is very difficult to represent three dimensions on a
two-dimensional sheet of paper or computer screen. You must drop
spikes to the 'floor' or 'origin' of the graph, otherwise it is impossible to
visualize the spread in the third dimension. It may be better to use a
series of two-dimensional scatterplots instead.
More than three variables: *use a series of two-, or three-, dimensional*
scatterplots, page 64.

25 (*Note*: the distinction here will be slightly fuzzy in some cases, but
essentially there are two basic types of test.)
The hypothesis is investigating differences and the null hypothesis   26
is that there is no difference between groups or between data and a
particular distribution [e.g. H_1 (alternative hypothesis) = white-eye
and carmine-eye flies have different mean development times, H_0
(null hypothesis) = white-eye and carmine-eye flies have the same
mean development time].
The hypothesis is investigating a relationship and the null   46
hypothesis is that there is no relationship [e.g. H_1 (alternative
hypothesis) = plant size is related to available phosphorous in the
soil, H_0 (null hypothesis) = plant size is not related to amount of
available phosphorus].

26 Data are collected as individual observations (e.g. height in   29
centimetres).
Data are in the form of frequencies (e.g. when carrying out a plant   27
cross and scoring the number of offspring of each type).

27 There are only two possibilities (e.g. white or pink).   28
There are more than two possibilities: *carry out a G-test, if your*
package supports it, page 72; otherwise use a chi-square goodness of fit,
page 75.
There are more than about eight possibilities: *a Kolmogorov–*
Smirnov test, page 86, may be more convenient than the chi-square
goodness of fit, page 75.

28 There are more than 200 observations in the sample: *carry out a*
G-test, page 72, if your package supports it; otherwise use a chi-square
goodness of fit, page 75.

There are 25–200 observations: *carry out a G-test if your package supports it, page 72; otherwise use a chi-square goodness of fit, page 75, but add a 'continuity correction' by adding 0.5 to the lower frequencies and subtracting 0.5 from the higher. This is very conservative and may result in a non-significant result when a marginally significant one is present (type I error). If your package supports the 'Williams' correction' then use that instead of the 'continuity correction'.*

There are fewer than 25 observations: *there are four possible solutions (listed in order of preference): use a binomial test if supported by your package; carry one out by hand if you are able; get a bigger sample; pretend you have 25 observations and use the instructions above.*

29 There is only one way of classifying the data (e.g. grouped by species). 30

There is more than one way of classifying the data (e.g. grouped by species *and* collection site). 38

30 There are only two groups (e.g. male and female or before and after). 31
There are more than two groups (e.g. samples from four different fields).
(*Note*: the null hypothesis is that all groups have the same mean so 35
if any two groups have different means you have to reject this null hypothesis.)
There are more than two groups and several measured variables [e.g. individuals divided by species (a grouping variable) and the measured variables are various anatomical characters or dimensions such as leaf length, stem thickness and petal length]: *canonical variate analysis, page 251.*

31 Two samples are 'paired'. This means that the same individual, 32
location or quadrat has been measured twice. This is the 'before-and-after' design (e.g. river nitrate level is measured at the same point before and after a storm).
Two samples are independent. There are different groups of 34
individuals in the two samples.

32 The data are normally distributed, there are at least 30 possible values and variances are, at least approximately, homogeneous: *carry out a paired t-test, page 92. To test for normal distribution use a Kolmogorov–Smirnov test, page 86, an Anderson–Darling test, page 89, a Shapiro–Wilk test, page 90, or a chi-square goodness of fit, page 75. A test for homogeneity of variance is often an option within the t-test in the package (e.g. a Levene test or Bartlett's test).*

A two-way ANOVA test is a potential alternative here but is more difficult to carry out than the paired t-test in most statistical packages, page 163 (use one factor of the ANOVA to represent 'before/after' and the other to represent the different individuals).
Above does not, or might not, apply. 33

33 Data have more than 20 possible values: *carry out a Wilcoxon signed ranks test, page 96.*
Data have 20 or fewer possible values (e.g. questionnaire results with a question of 'how do you feel' asked before and after exercise): *carry out a sign test if supported by your package (this is a very conservative but fairly low-power test), page 99. If this is not available in the package carry out a Wilcoxon signed ranks test, page 96.*

34 The data set is normally distributed, there are at least 30 possible values and variances are, at least approximately, homogeneous: *carry out a one-way ANOVA with one factor having two levels (one for each group), page 111, or use a t-test, page 103. To test for normal distribution use Kolmogorov–Smirnov tests, page 86, Anderson–Darling tests, page 89, a Shapiro–Wilk test, page 90, or chi-square goodness of fit, page 75. A test for homogeneity of variance is often an option within the t-test or the ANOVA in the package (e.g. a Levene test).*
The traditional method is to use a t-test for this type of experiment but it is no better than ANOVA in this circumstance as both tests give an identical result, although many packages have versions of the t-test that make an adjustment to the degrees of freedom to account for violations of the assumptions of the test.
The data set does not, or might not, fulfil the conditions above: *carry out a Mann–Whitney U test, page 119 (sometimes called Wilcoxon–Mann–Whitney or Wilcoxon two-sample test; not a Wilcoxon signed ranks test). (The Kruskal–Wallis test is an alternative but is less powerful.)*

35 Samples are 'repeated measures'. This means that the same individual or location is measured through time. This is an extended 'before-and-after' design (e.g. lake turbidity is measured at the same point each year for several years). 36
Each sample is independent. There are different groups of individuals in each samples. [It is important that no individual is present more than once in the data set, otherwise problems (of inappropriate replication) reduce the power of the statistical test.] 37

36 The data for each factor combination are normally distributed, there are at least 30 possible values and variances are, at least approximately, homogeneous: *carry out a two-way, repeated-measures* ANOVA *with one factor having a different level for each sampling repeat and a second factor having a level for each individual you are sampling, page 127 (easy if you have only five rivers measured each year but very tedious to input and difficult to interpret if you have 50). Be aware that if your package does not support repeated-measures designs the degrees of freedom in a two-way* ANOVA *should be reduced to compensate for the design. To test for normal distributions you can use Kolmogorov–Smirnov tests, page 86, Anderson–Darling tests, page 89, a Shapiro–Wilk test, page 90, or a chi-square goodness of fit, page 75, although in practice it is usual to use experience to determine whether the data are likely to be normally distributed. Furthermore,* ANOVA *is quite robust to small departures from a normal distribution.*

The data set does not conform to the restrictions above and you only have one observation for each repeat of each sample: *carry out a Friedman test with one factor having a different level for each sampling repeat event, page 123, (e.g. before, during, after) and one factor having a different level for each individual (e.g. person) you are sampling.*

Neither of the above apply. This is difficult! It often results from poor planning of the experiment: *usually it is best to carry out an* ANOVA, *page 163, as if the data conformed to the assumptions of distribution and variances but to treat the resulting P-values with caution, especially if a P-value is between 0.1 and 0.01.*

37 The data for each factor level are normally distributed, there are at least 30 possible values and variances are, at least approximately, homogeneous: *carry out a one-way* ANOVA *with the one factor having one level for each group, page 129. (Note: the t-test can only be used on two groups.) If the result is significant then you need to carry out a* post hoc *test to determine which factor levels are significantly different from which. If you are cautious, or unsure, use a Kruskal–Wallis test instead, page 142.*

The data set does not, or might not, fulfil the conditions above: *carry out a Kruskal–Wallis test with one factor having a level for each group, page 142. (Note: the Mann–Whitney U test only works for two groups so is not appropriate here.)*

38 There are only two factors/ways of classifying the data (e.g. strain and location).   39

There are three factors/ways of classifying the data (e.g. sex, region and year).   43

There are more than three factors: *use the key as if there are three* (43)
factors and extrapolate. Multifactorial experimental designs become
increasingly difficult to interpret because there are so many possible
interactions between factors and it is often easiest to leave out factors
that you have proved to have no significant effect, page 182.

39 There is no replication (i.e. only one value assigned to each   40
combination of the two factor levels) (e.g. the basal trunk
diameters after 2 years are collected from four strains of apple tree
grown under four watering regimes but with only one tree under
each watering condition).
There is replication (i.e. there are two or more values for each   41
combination of the two factors).

40 The data are likely to be normally distributed within each
factor combination (it is impossible to test this when there is
only one observation in each factor combination). Data such as
lengths and concentrations are likely to be appropriate but
judgement is required: *carry out a two-way ANOVA, page 152, but*
note that you will not be able to look for any interaction between
the two factors.
You are cautious, or have a data set that is unlikely to be normally
distributed: *carry out a Friedman test, page 146, although be warned*
that this test is quite weak.

41 Factors are fully independent of each other.   42
One factor is 'nested' within another (e.g. if there are three
branches sampled from each of three trees then branch is said to be
'nested' within trees): *carry out a nested ANOVA, page 193 (a.k.a.*
hierarchical ANOVA). (Note: there is no non-parametric equivalent (i.e.
one that makes fewer assumptions about the distribution of the data)
of this test.)

42 The data set is normally distributed within each factor
combination, there are at least 30 possible values and variances are
approximately equal: *two-way ANOVA, page 163, measure of the*
interaction between the two factors is possible.
The data set is not as above. Versions of a non-parametric, but
low-power, equivalent of a two-way ANOVA making fewer
assumptions about the data (i.e. non-parametric) are a fairly recent
innovation and are not yet appearing in statistical packages. If the
experiment is balanced, or nearly so (i.e. there are the same number
of observations for each combination of factor levels): *carry out a*
Scheirer–Ray–Hare test, page 175. This will, almost certainly, not be in
your statistical package but can still be carried out with a little
modification of other tests. See the section describing the test for details.

43 (*Note*: there are no non-parametric tests available from here on so if the data set does not fit the assumptions of the test you have no alternatives. ANOVA is quite robust to failure to meet its assumptions but be aware, especially if results are close to significance thresholds.)

All factors (ways of grouping the data) are independent of each other. 44

At least one factor is nested in another (e.g. in an experiment the variable is blood sugar level in mice. The factors are litter, female and food provided. If there are two litters from each of two females then litter will be 'nested' within female. Neither litter nor female will be 'nested' within food). 45

44 There is only one observation for each combination of factor levels: *carry out a three-way* ANOVA, *page 183. You will not be able to calculate the significance of the three-way interaction but you will be able to do this for the interaction between each combination of two factors. (Note that any main factors that prove to be non-significant can be left out of the analysis to reduce the complexity of the design.)*

There is more than one observation for each combination of factor levels: *carry out a three-way* ANOVA, *interaction terms are possible, page 184.*

45 One factor is 'nested' within another the third is independent (as in the mouse example in 43): *carry out an* ANOVA *involving both hierarchical and crossed factors, page 192. This is often difficult to reach in statistics packages although the design is a common one. If you only have one observation for each combination of factor levels then an interaction term cannot be tested (this is because it has to be used as the residual or error term).*

One factor is 'nested' within another that is itself 'nested' within a third (e.g. in a water pollution survey the variable is nitrate concentration. Several samples have been taken from five streams from each of three river systems and this has been done in two countries. The factors are stream, river and country. Stream is nested within river and river nested within country): *carry out a nested or hierarchical* ANOVA, *page 193.*

46 (The choice you have here is one that is frequently confused: be careful.)

The purpose of the test is to look for an association between variables (e.g. is there an association between wing length and thorax length?). You have not set (controlled) one of the variables in the experiment. There is no reason to assume a 'cause-and-effect' relationship. This is a test of correlation. 47

One or more of the variables has been set (controlled or selected) by the experiment or there is a probable 'cause' and 'effect' or functional relationship between variables. One of the uses of regression statistics you are moving to is prediction (e.g. the experiment is looking at the effect of temperature on heart rate in *Daphnia*. You are expecting that heart rate is affected by temperature but wish to discover the form of the relationship so that predictions can be made). This is a regression type of test.   53

47 Data are in the form of frequencies (e.g. number of white flowers and orange flowers).   48
There is a value for each observation. Variables should be paired etc. (e.g. an observation of two variables, cell count and lung capacity, from one individual).   50

48 There are two variables: *if you follow this thread further you will reach tests that are often awkward to carry out in packages and are often easier to calculate by hand. If you do calculate them by hand you may have to look up the significance level using a χ^2 table.*   49
There are more than two variables: *simultaneous comparisons of frequencies for more than two classifications are very difficult to interpret. It is best to compare them pairwise.*

49 The two variables each have two possible values (e.g. yes/no or male/female): *calculate a phi coefficient for a 2×2 table, page 209, if your package supports it or you can do it by hand. This test is a special case of a contingency chi-square calculation, page 199.*
At least one of the variables has more than two possible values (e.g. a crude land classification, forest/scrub/pasture/arable, is compared to an estimate of the density of a small mammal: common/rare/absent): *calculate a contingency chi-square, page 199, and, if your statistical package supports it, a Cramér coefficient, page 208.*

50 There are two variables.   51
There are more than two variables.   52

51 Both sets of data are continuous (have more than 30 values) and are approximately normally distributed (a good way to get a feel for this is to produce a scatterplot which should produce a circle or ellipse of points): *carry out a Pearson's product-moment correlation, page 210 (coefficient is called r). This is the standard correlation method.*
Data are discrete, or not normally distributed, or you are unsure: *use a Spearman's rank-order correlation coefficient, page 214, or a Kendall rank-order correlation coefficient, page 218 The marginal advantage of the former is that it is slightly easier to compare with the Pearson product-moment correlation while the latter can be used in partial correlations.*

Data are ranked: *use a Kendall rank-order correlation coefficient, page 218. (The Spearman's correlation is marginally inferior in this case.)*

52 (*Note*: partial and multiple correlations are difficult to interpret.)
All sets of data are continuous and approximately normally distributed, and you are interested in the direct level of association between pairs of variables: *use pairwise measures of association using a Pearson's correlation, page 210.*
All sets of data are continuous and approximately normally distributed, and you are interested in the overall pattern of association: *use partial correlation, page 237, which looks at the correlation between two variables while the others are held constant. (Multiple correlation is a possibility but is rarely supported in packages. Its disadvantage is in interpretation and its inability to distinguish positive and negative relationships.)*
Above do not apply, or you are cautious: *carry out Kendall partial rank-order correlation coefficient, page 237, a test that finds the correlation between two variables while a third is held constant. This may not be supported by your package. If it is not, pairwise testing is the only alternative.*

53 The dependent variable is discrete, or not normally distributed or   54
ranked. Be warned that non-parametric regression is required and that this is rarely available in a statistical package.
The dependent, or 'effect', variable is continuous and at least   55
approximately normally distributed with the same variation in 'effect' for any given value of the 'cause' variable. [There will often be a requirement for a transformation of the data. Proportions and percentages can be transformed using the arcsine transformation (page 44) or probits. Other distributions may be normalized using reciprocal transformations or many other possibilities. It is important that efforts are made to fulfil the requirements for approximately normal data with equal variance using transformations.]
The dependent 'effect' variable is a proportion or frequency (e.g. proportion of population with a disease). The 'cause' variable is measured without error and chosen or set by the experimenter: *use logistic regression, page 230.*

54 There is one independent 'cause' variable and one dependent   (55)
'effect' variable: *use Kendall robust line-fit method. If this is not available consider reframing (usually by simplifying) your hypothesis somewhat to fit a non-parametric correlation. The only other alternative is to continue to a parametric test (55), being very cautious with interpretation of the results.*

All other designs: *there is no satisfactory non-parametric test and* (55)
certainly nothing in a statistical package yet. Either reframe the
hypothesis or go to 55 and continue with a parametric test. If there are
two 'cause' variables and one 'effect' then the 'cause' variables might be
divided into a small number of categories (e.g. low, medium and high)
and then a Scheirer–Ray–Hare test could be carried out, page 175.

55 There is one dependent variable ('effect') and one independent   56
    variable ('cause').
    There is one or more dependent variable ('effect') and two or more   58
    independent variables.
    The data for the dependent variable can be classified into more
    than one group (e.g. by species or sex). There is a variable that may
    affect the dependent variable: *analysis of covariance (ANCOVA) is*
    required, page 238. This is a technique where the confounding
    variable, known as the covariate, is factored out by the analysis
    allowing comparison of the groups. Complex designs are possible but
    the most common is analogous to a one-way ANOVA with the data (e.g.
    dry weight) in classes (e.g. cultivars) and a variable known to be
    confounding factored out as the covariate (e.g. degree days).

56 The independent 'cause' variable is measured without error.   57
    There is known to be some measurement error associated with the
    independent variable: *a model II regression is required, page 235, or*
    Kendall robust line-fit method, page 230. This is a rarely used
    technique and only occasionally appears in statistical packages. It has
    the odd property of always overestimating the slope of the relationship
    compared to the result from a normal (model I) regression, page 221.

57 (As the theoretical shape of the relationship is often unknown the
    usual strategy here is to try both methods and see which gives the
    better fit.)
    The relationship is likely to be a straight line or you are not sure of
    the form of the relationship: *linear regression, page 221 (a.k.a.*
    model I regression). [Note: in many cases the independent variable can
    be transformed to straighten the relationship between cause and effect
    (e.g. if the independent variable is size and is right-skewed then a log
    transformation will often improve a linear fit).]
    The relationship is curvilinear or complex: *polynomial regression or*
    quadratic regression (a special case of polynomial regression), page 235.

58 There is one dependent 'effect' variable and two or more   59
    independent 'cause' variables.
    There are several 'cause' and 'effect' variables: *use path analysis,*
    page 243.

59 Your primary aim is to find the 'cause' variable(s) that are the best predictors of the 'effect' variable: *use stepwise regression, page 242.* You want to establish a model using all available 'cause' variables: *use multiple regression, page 242.*
(The distinction between these two is rather arbitrary.)

60 You have arrived at principal component analysis, discriminant function analysis and other multivariate techniques for exploring your data. The usual result of this type of exploration is to identify simple relationships hidden in the mass of the data. Some of these tests are described in Chapter 9.
There are several observed variables that are approximately continuous and you have no preconceived notion about division into groups: *use principal component analysis, page 244.*
There are a variety of variables that may be a combination of 'causes' and 'effects': *use path analysis, page 244.*
There are two or more sets of observations and one or more grouping variables: *use multivariate analysis of variance (MANOVA), page 256.*
There are two or more sets of observations, one or more grouping variables and a recorded variable that is known to affect the observed variables (e.g. temperature): *use multivariate analysis of covariance (MANCOVA), page 259.*
There are several observed variables for each individual that are approximately continuous and individuals have already been assigned to groups (e.g. species): *use canonical variate analysis, page 251.*
There are several observed variables for each individual that are approximately continuous, individuals have already been assigned to groups (e.g. species) and the intention is to assign further individuals to appropriate groups: *use discriminant function analysis, page 251.*
There are several observed variables for each individual that are categorical or nominal, individuals have already been assigned to groups (e.g. species) and the intention is to assign further individuals to appropriate groups: *use logistic regression, page 230.*
There are several observed variables for each individual and you wish to determine which individuals are most similar to which: *use cluster analysis, page 259.*
You have data on the relative abundance of species from various sites and wish to determine similarities between sites: *use cluster analysis, page 259, or TWINSPAN, page 263.*

Hypothesis testing, sampling and experimental design

4

This chapter expands on some of the ideas introduced in Chapter 2.

Hypothesis testing

Much of scientific investigation is based on the idea of hypothesis testing. The idea is that you formalize a *hypothesis* (H_1) into a statement such as 'male and female shrimps are different sizes', collect appropriate data and then use statistics to determine whether the hypothesis is true or not.

However, it is not quite as simple as that. The statistical tests do not give a simple answer of true or not. First you have to realize that every hypothesis will have an associated *null hypothesis* (H_0) and most statistical tests use the null hypothesis as a starting point.

So, for this example, the hypothesis (H_1) is 'male and female shrimps are different sizes' and the associated null hypothesis (H_0) is 'male and female shrimps are *not* different sizes'.

What a statistical test determines is the probability that the null hypothesis is true (called the *P*-value). If the probability is low then the null hypothesis is rejected and the original hypothesis accepted.

Acceptable errors

In reality, the null hypothesis is either true or false. Unfortunately, we only have a sample of all the individuals in a population and the statistical test only gives an indication of how likely it is that the null hypothesis is true based on the sample available. There are two ways of making the wrong inference from the test. These two types of error are usually called type I and type II errors.

Choosing and Using Statistics: A Biologist's Guide, 3rd Edition. By Calvin Dytham.
Published 2011 by Blackwell Publishing Ltd.

		Null hypothesis	
		Accepted	Rejected
Null hypothesis	True	Correct	**Type I error**
	False	**Type II error**	Correct

In a *type I error* the null hypothesis is really true (male and female shrimps are not different sizes) but the statistical test has led you to believe that it is false (there is a difference in size). This type of error is potentially very dangerous and could be seen as a 'false positive'.

In a *type II error* the null hypothesis is really false (male and females are really different sizes) but the test has not picked up this difference. Small sample sizes will often lead to a type II error. This type of error is less dangerous than the type I but should still be avoided if possible.

The ideal statistical test should have an equal, and hopefully very low, chance of the two types of errors. A test which increases the chance of getting a type II error while decreasing the chance of a type I is said to be 'conservative' while one that increases the chance of a type I error is said to be 'liberal'. Although it is best to achieve this balance of type I and type II errors, a cautious approach is to err towards more 'conservative' tests.

P-values

Errors are the inevitable consequence of results based on probability. The lower the probability (*P*-value) the more confident you can be in the rejection of the null hypothesis but you can never be totally sure, unless you have measured the whole population, that you are correct. It is a usual convention in biology to use a critical *P*-value of 0.05 (often called alpha, α). This means that the probability of observing data as extreme as this if the null hypothesis is true is 0.05 (5% or 1 in 20); in other words, it indicates that the null hypothesis is unlikely to be true. In biological sciences it is convention that whenever a statistical test gives a result with a *P*-value less than 0.05 we reject the null hypothesis and accept the alternative hypothesis.

There is nothing magical about $P < 0.05$, it is just a convention. If you use a lower critical *P*-value then the chance of making a type II error is increased. If you choose a higher critical *P*-value then you increase the chance of making a type I error.

It is worth pointing out that if a *P*-value is less than 0.05 it does not *prove* that the null hypothesis is false, it just indicates that it is unlikely to be true. Indeed statistics can never *prove* anything, it can only suggest that a hypothesis is very likely to be true or untrue. It is also worth noting that a *P*-value above 0.05

certainly doesn't prove that the null hypothesis is true: it just indicates that there is not enough evidence to reject it.

It is conventional to indicate degrees of significance using asterisks in tables, or sometimes on figures. A single asterisk is usually used for P-values between 0.05 and 0.01, two asterisks for values below 0.01 and then three asterisks for results below either 0.005 or 0.001. If asterisks are used in this way they should be explained in a figure or table caption.

One final point about P-values is that when more than one test is used the critical P-value used should be reduced to retain a critical level of 0.05 across an experiment. This makes good sense as any experiment including 20 statistical tests should, on average, generate one significant result even when there is no biological effect. The two main methods for adjusting P-values to retain the experiment-wide threshold are discussed in more detail in the section on correlation (page 199).

Sampling

Nearly all statistical tests make a fundamental assumption that sampling of individuals will be at random from all the individuals that could possibly be sampled. This sounds simple but achieving a random sample is not always easy. In almost all biological studies it will be impossible to account for every individual in a population. Therefore it is necessary to examine a subgroup of the total population and extrapolate from this to the whole population. The process by which the subgroup of the population is selected is called sampling.

If a population is evenly distributed through a habitat then a single small sample would be enough to gain a good estimate of whatever it is you are interested in (e.g. total population size, mean age or weight). However, this is rarely the case and most populations have distributions that are either random or clumped. In such populations a single sample is unlikely to produce a good estimate of population size or mean height or the variance of leaf thickness.

There are a wide variety of sampling strategies in use. It is important to choose a strategy that is appropriate to the population being investigated. There are several steps in the development of a sampling strategy, as described below.

Choice of sample unit

A sampling unit may either be defined arbitrarily, such as a quadrat, transect or pitfall trap, or be defined naturally, such as leaf or individual. Usually, naturally defined sample units will be obvious but the choice of an arbitrary unit size may be important. If the unit is a quadrat then there will obviously be a trade-off in effort between the number of sample units that can be observed and their size, simply because it takes more time to get information from a large quadrat.

A sample unit might also be a length of time (e.g. if you are investigating pollination then number of times a flower is visited in a series of time periods of set length might be your data set).

> The size of sample unit will usually remain constant but may sometimes be variable, especially if the characteristics of a population distribution are being investigated. Methods using variable sample unit size to investigate distribution pattern are called quadrat-variance methods. These methods allow an observer to gain an insight into patterns of distribution in space or time by analysing the different characteristics of the samples (e.g. mean and variance) using different sample units.

Number of sample units

This is nearly always determined by the amount of labour available: the more time and people that are available the more information can be collected. However, it is possible to calculate the number of sample units required to produce an accurate estimate of the population size. In general more sample units will be preferred as the number of sample units increases the accuracy of statistical tests. However, quantity should not be increased at the expense of quality. Poor-quality data will have more inherent error and therefore make the statistics less powerful.

If you require general advice on the number of observations to make then I can only suggest that, as a rule of thumb, you need at least 20 observations for a sample using a measured variable and many more than that if the variable is a simple categorical one.

Positioning of sample units to achieve a random sample

An unbiased estimate of a population is only possible if the sample units are representative of the total population. The easiest way of achieving this is for each sample unit to contain a random sample of the population under investigation. If quadrats are being used then their position within the area under investigation should be chosen using random numbers to generate two co-ordinates that are then used to position a corner of the quadrat. Although this method of choosing a position using random numbers often requires an area to be marked out, it is to be preferred over the quasi-random techniques, such as throwing a quadrat, that are certain to introduce some involuntary observer bias.

Selecting random individuals in an area for study can be difficult. Imagine a typical scheme for locating random plants: random coordinates are chosen and the nearest plant to the random location is selected for study. This apparently makes a random choice of individual plants, but in reality it introduces bias as isolated plants are much more likely to be selected than plants in the middle of clumps. Indeed the only way to really select individuals at random is, rather impractically, to label and number every individual and then select randomly from that list.

Random walks are another way to sample at random without requiring the area to be marked out. The observer walks a number of paces determined by a random number and then makes an observation or places a quadrat. Then another random walk is taken before the next observation, and so on. The advantages of this method are that sampling can be very rapid and that it requires little preparation. The drawback is that this type of sampling may be severely biased by the observer.

True random samples are ideal in a perfectly homogeneous habitat, but in a heterogeneous habitat they are likely to produce a biased sample with an estimated variance greater than that of the total population. A simple method used to minimize this problem is to take a *stratified random sample*. The method is simple: the total area is divided into equal plots and an even number of sample units is taken at random from each plot. It is possible to divided the total area into plots of different sizes if there are known to be different habitats in the total area. In this second case the number of sample units from each plot should be proportional to its size.

It might be tempting to conduct systematic sampling with sample units placed at regular intervals across a study area. There is a statistical problem with this strategy as most statistics require that a sample is taken at random from a population. However, some field ecologists suggest that estimates derived from systematic sampling are, on average, better than those from random sampling.

Timing of sampling

Most populations will be affected by season, time of day and local weather conditions. It is very important that timing is taken into account either by sampling strategy or by later analysis.

What I have been considering here is the problem faced by an ecologist working in the field and trying to design a suitable sampling strategy. The use of the very powerful statistical technique analysis of variance (ANOVA) is more common in the situation of a controlled experiment where you are analysing the effects of different levels of a treatment (e.g. concentrations of fertilizer or temperature) on some measured aspect of a population. Then, to get a true estimate of the effect of the treatment, experimental design will be of paramount importance.

Experimental design

I do not intend to say very much about experimental design here as there are whole books dedicated to the subject. However, that should not imply that experimental design is an uninteresting or unimportant subject. The appropriate design of an experiment is the key to successful analysis of a problem for without the correct design you will never have the right sort of data. The problems of sampling still exist in a designed experiment but the control of the system allows the experimenter to ensure that there are sufficient individuals to sample, and that all factor combinations have the same number of observations.

Control

The idea of control in experiments is to remove the effect of all other factors apart from the one that is being investigated. The word is often applied to a group that has not been altered by an experimental manipulation and is used to compare with a group that has. The assumption is that everything apart from the manipulation is the same in the two groups so any differences must be the result of the manipulation. There are different ways of applying controls and different types of control. It is important to consider whether the control used in an experiment is adequate to convince a sceptical reader that the effect 'proved' by the statistics is real or not. As a general rule of thumb more control is required! It is always tempting to focus on the more interesting manipulated groups and not give enough attention to control.

Procedural controls

These are often overlooked in experimental designs. The idea is that everything done to the manipulated group, apart from the actual treatment, is done to a procedural control group that is the same size as the experimental group and the untouched control group. The idea of procedural control groups has been widely used in recent medical studies and shown some interesting results. For example, a new supplementary treatment for a common disease is being investigated. Everyone in the study is given the conventional treatment. Each individual is then randomly (and secretly) assigned to one of three categories: the control group is given nothing else, the experimental group is told about the supplementary treatment and given the new drug and the procedural control group is treated exactly as the experimental group except they are given a dummy drug (e.g. chalk tablet or water injection). In this way the effect of the drug can be differentiated from the effect of the procedure.

Procedural controls are especially important if the experimental system requires a lot of preparation through building exclosure fences or with repeated visits to a site or many interventions in a laboratory population. The technique should nearly always be used in conjunction with the 'untouched' control.

Temporal control

This is another aspect of experimental design that is worth incorporating. If the effect of a long-term manipulation is to be considered and there is only one control group and one experimental group available it is better to start the manipulation *after* the monitoring process has been underway for some time. The reason for this is that the differences between the two populations *without* any manipulation should be accounted for before the differences following the manipulation are tested. The ideal experiment will use half of the time for before and half for after manipulation. For example, if there are two lakes available to study the effect of eutrophication (surplus of nutrients) then the best design for a 2-year study is to monitor the lakes untouched for the first year and then to add nutrients to one of the two lakes during the second year.

Experimental control

This is any control of environment imposed by the experimenter. This is the classical type of control and is properly employed to remove all possible effects on the observations other than that from the experimental manipulation itself. The best advice is always to control as many factors as possible. So if the effect of CO_2 levels on plant growth is being investigated then the experiment should control all the factors that may affect growth: light, temperature, humidity, water availability, soil organisms, soil type and nutrient availability. The degree of environmental control required to isolate the effect of the one factor being manipulated often leads to a very artificial situation with organisms being kept in isolation in perfect conditions. These controlled environments are often so far removed from the real world that the results are not really very informative.

Experimental control can be very expensive, requiring growth cabinets, controlled-temperature chambers or incubators for even rather simple investigations.

Statistical control

This is an alternative to experimental control. Rather than fixing all the possible factors that can affect the observations the factors are measured instead. Careful recording of all the environmental conditions, both biotic and abiotic, that are known to affect the observations being collected can then be used in statistical analysis of the data. Providing the experiment is not confounded (e.g. if all the manipulated individuals are in a cold area and all the unmanipulated in the warm) it is often possible to unpick the various effects and remove them from the analysis to leave only the effect of the manipulation. If statistical control is to be attempted then efforts should be made to ensure that adequate monitoring of all the possible effects is carried out and that the individuals in experimental and control groups experience a range of conditions.

Statistical control is usually cheaper than experimental control but requires more effort on the part of the researcher.

Some standard experimental designs

The *Latin square* is a system for placing replicates of treatments so that each of the treatment levels experiences each column and row of the experimental area. The reason for doing is to avoid confounding the effect of the experimental treatment with any other factor that might be present in the experimental area (e.g. a gradient of soil quality). The arrangements for the four treatments suggested here is just one of many possible arrangements. Any arrangement of treatments such that each appears once in each column and row is OK although it is probably best, as here, to have each treatment level only occurring in one corner.

A	B	C	D
C	A	D	B
D	C	B	A
B	D	A	C

If the experiment is carried out in a series of locations (often called blocks in statistical jargon) it is important to ensure that each of the treatment levels is equally represented in each of the blocks, otherwise any difference in conditions will be confounded with the treatment levels being investigated. Furthermore the position of the treatment levels within the blocks must not be repeated.

Block 1

A	C
D	B

Block 2

B	A
C	D

Block 3

D	B
A	C

Block 4

C	D
B	A

If a large number of samples are to be assigned to different treatment levels there are third obvious ways of assigning the levels: first to do all of level one then two and so on, second to carry out the assignments entirely at random and third to keep cycling through the levels in sequence. Each of these has problems. The first method will confound any external changes with the different treatment levels (e.g. if the experimenter becomes more efficient during the process). The second method is appealing but often leads to unwanted runs of the same treatment or too few replicates of particular treatments. The third may also confound the treatment levels with an external influence. The best strategy is a combination of the second and third methods and is called *stratified random* assignment. The assignments are made in batches with each treatment level appearing an equal number of times in the batch (usually one or two) but assigned at random. In the example shown there are three treatment levels (X, Y and Z) assigned twice each in five batches of six.

Batch 1	Batch 2	Batch 3	Batch 4	Batch 5
X	Z	Y	X	Z
Z	X	Y	Z	X
X	Y	Z	Y	X
Y	X	X	Z	Z
Y	Z	Z	Y	Y
Z	Y	X	X	Y

5 Statistics, variables and distributions

There are many books available that discuss the history, philosophy and workings of statistics at length. That is not my purpose here, but it is important to have at least some idea of what statistics are, how different statistics are appropriate in different circumstances and that there are different types of data that you might collect. This chapter covers much of the same ground as Chapter 2 but in much more detail. However, I'm still only scratching the surface here and this section should only be used as a set of notes or pointers to further investigation of these subjects.

What are statistics?

In biology we are often concerned with groups of individuals. These 'individuals' might be single insects but they could also be, for example, herds, or species, or blood cells. In most cases it is totally impractical to measure every individual in the group or groups we are interested in. What we are forced to do instead is to take measurements from a subset of the group. We call these subsets of the whole group *samples*.

We can ask and answer questions about the groups by formulating hypotheses. A simple question could be 'is species A bigger than species B?'. If we had access to data from all individuals in a group we could answer this type of question very easily. However, we only have the sample and from the sample we have to extrapolate to the whole group. This is the job of statistics.

For example, if the hypothesis is that the mean sizes of populations of pike in two lakes are different we could easily find the answer if we had measured all the fish. However, in reality we only have a sample of 20 fish from each lake and the means of those samples might not be the same as that of all the fish. We carry out statistics on the information we have in the two samples to determine the probability that the hypothesis, or more usually its associated null hypothesis, is true. The idea of hypothesis formulation and testing has already been discussed in Chapter 4.

Choosing and Using Statistics: A Biologist's Guide, 3rd Edition. By Calvin Dytham.
Published 2011 by Blackwell Publishing Ltd.

Types of statistics

I intend to say as little as possible about types of statistics here. However, I feel it is important to give a feel for the differences and the way that statistics have been traditionally divided.

Descriptive statistics

These are usually the first to be calculated. They give information about the data you have collected. This can be a measure of the 'position' of the data – that is, mean or median – and the 'dispersion' – or how variable the data are. Some descriptive statistics, such as means (averages), will be familiar to everyone.

There is a division of statistics into two groups that are usually labelled 'parametric' and 'non-parametric'. This distinction is very real for statisticians but for those of us just using the tests it seems rather artificial.

Parametric statistics

These statistics make assumptions about the form of the data under investigation. For instance, they usually require variables to follow known distributions, usually the normal. If the data do conform to the assumptions then these tests are usually more powerful and should therefore be preferred. There are also types of questions that can only be answered if assumptions about distributions are made.

Non-parametric statistics

These are statistics that require little or no knowledge of the distribution of the data. Therefore they are often called 'distribution-free', 'ranked' or 'ranking' tests. In general these tests are less powerful but 'safer' if you have not tested all the assumptions for a parametric test. Non-parametric tests are also somewhat restrictive and cannot be used to answer some more complicated questions.

In this book, unlike many other books, the chapters are not ordered according to the type of statistics. I have used the type of question you want to ask as the method of dividing up the book.

What is a variable?

To carry out any statistics you need some data to work with. First you decide what it is you are interested in and then select a suitable variable. The variable is the property that you measure. It is the food of the statistics and choosing variables is something you must get right. For example, if you are interested in the occurrence of a scale insect on two strains of citrus trees then a suitable variable might be 'number of scale insects on a leaf'. However, if the strains differ in leaf density or size of leaf then a more appropriate variable could be 'number of scale insects per square centimetre of leaf'. Then the insects may be bigger on one strain than the other so perhaps 'mass of scale insects per square centimetre of leaf'. Each time the variable is refined in this example it becomes more difficult to obtain, taking more time and effort. There is a trade-off. More effort required for each observation leads to fewer data in total so any refinements to the variable collected must be warranted.

Choosing the best variable is something of an art.

It is important to ensure that the variable or variables you choose to measure or collect are appropriate to the task.

> *Note*: I use the term variable throughout this book as it is the one in common usage although the correct term is variate.

Types of variables or scales of measurement

There are many types of variable.

Measurement variables

These are variables where a numerical value is assigned. They can be further subdivided.

Continuous variables

This type of variable (sometimes called 'interval' variables) theoretically has an infinite number of values between any two points. Of course in practice the accuracy of measurement will not be perfect, as it will be limited by the observer and the equipment used. Therefore there will only be a limited number of possible values between any two points. Obvious examples of continuous variables are lengths, weights and areas.

> *Note*: accuracy and precision are two words that are often confused.
>
> **Accuracy** is the closeness to the real value. This is usually set by the observer or the equipment and should be chosen as appropriate to the variable. When you write down a value it should reflect the accuracy with which the measurement was taken. If you measure to the nearest 0.1 g then 5 g should be written as 5.0 g, not 5.00 g;
>
> **Precision** is the closeness of repeated measures to the same value. It is possible to have data that are very precise but very inaccurate. For example, your balance gives exactly the same value for repeated measures of the same object but they are all overweight because the balance was not calibrated properly. The data obtained would be precise but inaccurate.

Discrete variables

Unlike continuous variables this type of variable (also called 'discontinuous' or occasionally 'meristic') has a limited number of possible values. These possibilities are often, but not always, integers. For example, number of live-born offspring in a litter of mice can only ever be an integer as there is no possibility of recording a fraction of an offspring.

Discrete variables are often produced by questionnaires. Respondents are offered choices such as: 1, strongly disagree; 2, slightly disagree; 3, neutral; 4, slightly agree; 5, strongly agree. There is clearly a continuous variable ('agreement') here and division of responses into categories in this way is rather arbitrary. It would be very easy to devise different ways of dividing the responses to obtain more or fewer possibilities.

The distinction between discrete and continuous variables can be rather blurred.

Example 1: a discrete variable becomes continuous. If you measure the number of cells in 1 ml of blood this must be an integer and therefore discrete. However, it has so many possible values that it is effectively continuous.

Example 2: a continuous variable becomes discrete. Seed diameter is a continuous variable but if you measure poppy seed diameter to the nearest 0.05mm there will be only a few possible values making it effectively discrete.

How accurate do I need to be?

It is often possible to use better equipment or become more careful when measuring to increase accuracy. However, increased accuracy will take longer and result in fewer data being collected: another trade-off. As a rule of thumb there should usually be between 30 and 300 possible intervals between the smallest and the largest value. If possible, adjust the accuracy of the measurement accordingly. Don't assume that measuring to as many decimal places as possible will make the data any better.

Ranked variables

When data are ordered by magnitude and exact values are not relevant the variable is called ranked. It is not assumed that the difference between 1 and 2 is the same as that between 3 and 4. Often it is possible to put observations into rank order without measuring at all. For example, plants from six pots could be ranked in 'health' order by simple observation and assigned values from 1 to 6.

Attributes

These variables (also called 'categorical' or 'nominal' variables) have few categories; usually 'yes' or 'no', 'male' and 'female' or a small number of possibilities.

For example, you could score flower colours as red, blue or yellow. Attributes or categories should not have any obvious sequence.

Derived variables

Derived variables (or 'computed' variables) are usually calculated from two (or more) other variables; for example, ratios, percentages, indices and rates.

> *Warning*: you lose accuracy by combining variables into ratios.
>   For example, if we round to 0.1 then 1.2 implies 1.15–1.25, giving a maximum error of 4.2%, and 1.8 implies 1.75–1.85 with a maximum error of 2.8%. If these two observations are combined into a single observation then 1.2/1.8 implies a range from 1.15/1.85– 1.25/1.75 or 0.622–0.714 giving a maximum error of 7%: this is much greater than the error in the original data.

Distributions of combined variables are often awkward. Be very careful with percentage data as percentages will often have rectangular ('flat' or 'uniform') distributions and/or have limits at 0 and 100%. However, percentages are a familiar and widely used method for expressing observations and there are a few statistical tricks available to help you deal with them.

Ratios can also lead to a loss of information. For example, both 5/10 and 500/1000 will give a ratio of 0.5, losing all information about the size of the sample.

Types of distribution

Why do you need to know about distributions? Just as there are different types of variable, there are different types of distribution. All parametric statistics and many non-parametric ones are based on features of distributions or on assumptions about data following certain distributions.

Discrete distributions

The Poisson distribution

This is a very useful tool to use as a starting point in many biological investigations. It is a distribution describing the number of times an event occurs in a unit of time or space. Usually a sample of time or space is taken and the number of events recorded. Examples of typical events are the number of fish-lice on a fish or number of influenza cases reported in a week.

The purpose of fitting to this distribution is to test for randomness or independence in either space or time. If the number of scale insects on leaves fits the Poisson distribution then it can be assumed that the assumptions hold. Therefore the occurrence of individual insects is unaffected by the presence of others, so we can infer that the scale insects arrive at random and that no leaf is 'full' of scale. If the distribution is significantly different from a Poisson then not all the assumptions hold and further investigation should follow.

Poisson distributions only require knowledge of the mean as mean and variance are equal. This property is also very useful as simple inspection of the mean and variance of observations in a sample will give you some idea of the form of a distribution.

If the variance is greater than mean then the population is more clumped (aggregated) than random. If the variance is less than the mean then it is more ordered (uniform) than random (see Fig. 5.1). Distributions may be described by simply quoting their variance/mean ratio, with a value of 1 indicating a random (Poisson) distribution, and higher values indicating clumping.

The binomial distribution

This is a discrete distribution of number of events. When there are two possible outcomes for each event the probability of each is constant. For example, if the probability of each birth producing a female is 0.5 (usually termed p) then the probability of a male is 1 minus 0.5 (also 0.5 in this case and often termed q) as there are no other possibilities. This means that each individual being born has a 50% chance of being female and 50% chance of being male.

If this is expanded to families with more than one offspring then we can start to apply probabilities to the proportions of males (M) and females (F). For example in a family with two offspring there are four possible outcomes: FF, FM, MF and MM (note that there are two routes to get one male and one female). As the chance of each event has already been determined as 0.5 then the chance of each of the four outcomes is 0.5×0.5 or 0.25. In other words there is a 25% chance of getting FF, 25% for MM and then 25% for each of MF and FM. So 50% of families with two offspring will have one of each sex.

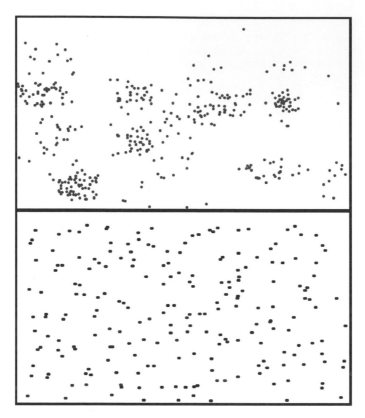

Fig. 5.1 Two hypothetical distributions of individuals in space. In the first the individuals are highly clumped or aggregated. If quadrats were used to sample from this population the variance in number of individuals per quadrat would exceed the mean. However, in the second distribution the individuals are more ordered than random and the results of number of individuals per quadrat would show a variance less than the mean.

This can be expanded further to three offspring where there are four possible families (reached through eight different sequences, each with a 0.125 probability of occurrence):

Female offspring	Male offspring	Probability	Sequences
3	0	0.125 (1/8)	FFF
2	1	0.375 (3/8)	MFF, FMF, FFM
1	2	0.375 (3/8)	MMF, MFM, FMM
0	3	0.125 (1/8)	MMM

There are many uses of this expansion from single events to groups in biological investigation. To stay with the male/female example for the moment, an investigation into 480 broods of song thrushes (*Turdus philomelos*) where there were five eggs surviving to fledging gave frequencies (numbers of observations) for each of the six possible categories of families:

Females	Males	Probability	Expected no.	Observed no.
5	0	0.03125	15	21
4	1	0.15625	75	76
3	2	0.31250	150	138
2	3	0.31250	150	142
1	4	0.15625	75	80
0	5	0.03125	15	23

The expected frequencies from the assumption of a binomial distribution can be tested against the observed numbers using a chi-square test or a G-test. In this case, despite having fewer broods with three of one sex and two of the other than was expected, the difference is not significant and therefore we accept the null hypothesis that the sexes of individuals in song thrush broods of five follow a binomial distribution with a p of 0.5 (i.e. there is a 50% chance of having a female offspring).

The binomial makes a very good starting place for a null hypothesis of even chances of events happening in all groups observed. If the binomial distribution were not followed then alternative explanations about aggregated or dispersed events have to be invoked.

The negative binomial distribution

In many organisms aggregation of individuals in time and/or space is almost ubiquitous. The negative binomial distribution is a discrete distribution that can be used to describe clumped data (i.e. when there are more very crowded and more sparse observations than a Poisson distribution with the same mean). There are reasonable assumptions that can be made about the way organisms distribute themselves that result in a negative binomial distribution. This allows a sensible null hypothesis about aggregated distributions to be made.

The hypergeometric distribution

This is another theoretical, discrete distribution that has some use in biology. The hypergeometric distribution is used to describe events where individuals are removed from a population and not replaced. It is therefore quite useful in small, closed populations that are being sampled destructively and also in the application of mark/recapture techniques.

Continuous distributions

The rectangular distribution

This distribution (also called a 'flat', 'even' or 'uniform' distribution) describes any distribution where all values are equally likely to occur. This distribution rarely appears in reality but it can sometimes be useful for generating a null hypothesis (see Chapter 4).

The normal distribution

This is the most important distribution in statistics and it is often assumed that data are distributed in this way. Therefore is it often important to determine whether the data set is a good fit to a normal distribution or not. Methods you can use to test this are the Kolmogorov–Smirnov test, the Anderson–Darling test, the Shapiro–Wilk test or a chi-square goodness of fit (see pages 75 and 86–92). These methods are not very sensitive when samples are small and should not be used if there are fewer than about 50 observations.

The normal distribution is a symmetrical, continuous distribution and is described by two parameters: the mean, μ (mu, describing the position), and the standard deviation, σ (sigma, describing the spread). These two parameters are estimated from samples and assigned the letters m and s.

The normal is sometimes called the Gaussian distribution. A normal distribution always has a characteristic bell shape.

In a perfect normal distribution (where μ is the mean and σ is the variance):

$\mu \pm \sigma$ contains 68.25% of the observations; 50% fall between $\mu \pm 0.674\sigma$;
$\mu \pm 2\sigma$ contains 95.45% of the observations; 95% fall between $\mu \pm 1.96\sigma$ (see Fig. 5.2);
$\mu \pm 3\sigma$ contains 99.73% of the observations; 99% fall between $\mu \pm 2.576\sigma$.

The standardized normal distribution

This is a derived distribution where each observation in a normal distribution is processed by subtracting the mean and dividing by the standard deviation. This gives a normal distribution with a mean of 0 and a variance of 1. The purpose of this transformation is to compare distributions that might have very different means on the same scale to look at the shapes of the distributions.

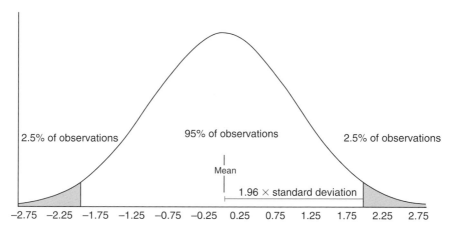

Fig. 5.2 In a normal distribution 95% of the observations will fall within 1.96 standard deviations of the mean. This leaves 2.5% of the observations in each of the tails of this symmetrical distribution (shaded).

Convergence of a Poisson distribution to a normal distribution

Even though a Poisson distribution is discrete (you can only get integers), when the mean number of observations is very large a Poisson distribution will approximate to a normal distribution. This could arise for example if you counted the number of springtails in a group of soil samples and found that they fitted to both Poisson and normal.

Note: binomial distribution with more than 100 observations (or fewer if $p \approx 0.5$) will also approximate to a normal distribution.

Sampling distributions and the 'central limit theorem'

The means of samples taken from any shape of parent distribution will themselves have a normal distribution: that is the central limit theorem. This is the basis for the rule that the standard deviation of the sample mean (i.e. standard error) of a sample is $s/\sqrt{n}$, where s is the standard deviation of the observations and n is the number of observations.

Describing the normal distribution further

Two types of departure from normality in a data set are skewness and kurtosis.

Skewness

This is another word for asymmetry; skewness means that one tail of the bell-shaped curve is drawn out more than the other (see Figs 5.3 and 5.4). Skews are either to the right or left depending on whether the right or left tails are

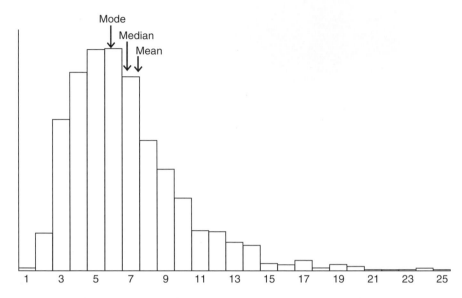

Fig. 5.3 This distribution is clearly right skewed and has a g_1 value well above zero. In a skewed distribution the mean is always nearer the tail than the mode with the median falling between the mean and the mode.

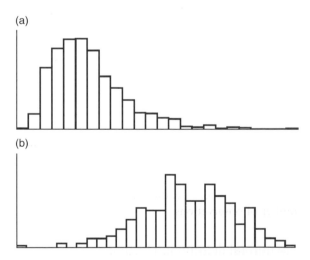

Fig. 5.4 These two frequency distributions are clearly not symmetrical. The data in (a) are right skewed and have a g_1 value of 1.53. The data (b) are left skewed and have a g_1 value of −0.335.

drawn out. (i.e. long right tail results in a right-skewed distribution). Statisticians label the true skewness parameter γ_1 (gamma$_1$) and the estimated value g_1. A negative g_1 indicates skewness to the left and a positive g_1 skewness to the right.

If a distribution is skewed the mean is nearer to the tail than the mode and the median, as shown in Fig. 5.3.

Kurtosis

This is a measure of the 'flatness' of a distribution. A symmetrical distribution can differ from the normal in being either *leptokurtic* or *platykurtic*. A *leptokurtic* distribution has more observations very close to the mean and in the tails. A *platykurtic* distribution has more observations in the 'shoulders' and fewer around the mean and tails. A bimodal distribution is, therefore, extremely *platykurtic*. The kurtosis parameter is γ_2 (gamma$_2$) and estimated by g_2.

In a perfect normal distribution both g_1 and g_2 are equal to zero. A negative g_2 indicates a platykurtic distribution and a positive g_2 leptokurtosis.

Is a distribution normal?

It is extremely unlikely that you will collect a data set that is perfectly normally distributed. What you need to know is whether the data set differs *significantly* from a normal distribution. One good way for checking data for departures from 'normality' is to use the *Kolmogorov–Smirnov* test, *Anderson–Darling* test or *Shapiro–Wilk* test. These tests compare two continuous distributions with the null hypothesis that they are the same (i.e. it tests the sample data against a normal distribution with the same mean and variance as the sample). All these tests are usually to be preferred over the chi-square goodness of fit method which is another commonly used method of determining whether data set is normally distributed. See Chapter 7 for details of the Kolmogorov–Smirnov, Anderson–Darling, Shapiro–Wilk and chi-square goodness of fit tests (pages 75–92).

Transformations

Parametric statistics assume that data set you are using is distributed normally. So first of all check that this is true using a statistical test fitting your distribution to a perfect normal distribution with the same mean and variance. If the data set is significantly different from normal try a transformation such as logarithmic, or square root, arcsine square root for percentage or proportion data, or probits or logits. There are many standard methods to try but as long as you treat each piece of data (datum) in exactly the same way you can do any transformation you like. Be warned that logarithmic (log) transformations require you to consider the base of the log. In some packages 'log' will give a base 10 log transformation, while in others it will give a natural log with base e (approximately 2.71).

An example

A study on feeding preferences in a marshland birds counted the number of grey herons (*Ardea cinerea*) seen in creeks and open water at different times of the day. The percentage in creeks is converted to a proportion and the angular transformation and logit transformations given.

Time	No. in creek	No. in open	Percentage in creek	Proportion in creek	Angular-transformed	Logit-transformed
0600	22	25	46.8	0.47	43.2	−0.13
1200	12	19	38.7	0.39	38.5	−0.46
1800	25	8	75.8	0.76	60.5	1.14

The angular transformation

The *angular* or arcsine square-root transformation is so routinely applied to percentage data that it warrants a description of the method. A percentage is converted to a proportion, the square root taken and then the **arcsine** (inverse sine or $\sin^{-1}$) is taken. To make sure you have the calculation correct, either use the values from the table above or check that 10% converts to about 18 after an angular transformation and 100% converts to 90. A common problem encountered with this transformation is that packages use radians rather than degrees and this must be accounted for. Remember that if you are converting direct from percentages rather than proportions, the variable to be converted should be divided by 100 as part of the transformation.

SPSS Assuming the percentages have been converted to proportions and are stored in a variable called 'prop', from the 'Transform' menu select 'Compute Variable'. In the 'Compute Variable' box that appears type a name for the target variable (say, 'angular'). Select 'All' in the 'Function group:' list. Then from the list of 'Functions and Special Variables', select 'Arsin' and click the up arrow to add it to the 'numeric expression'. Next, with the question mark highlighted in blue, select 'Sqrt' from the functions list. Finally, with the question mark selected again, select the variable 'prop' and move it across to the 'numeric expression' (either by double clicking, or by highlighting and clicking the right arrow). Finally, the correction for converting radians to degrees needs to be applied and the expression multiplied by 57.295.

The 'numeric expression' should now read '57.295*ARSIN(SQRT(prop))'. Click 'OK' and the converted numbers should appear in a new variable.

R The sin^{-1} function is 'asin()' and the square-root function is 'sqrt()'. Assuming you have percentages in a variable labelled as 'x' and you want the results in a variable labelled 'angularx' type in the following:

```
> angularx=57.295*asin(sqrt(x/100))
```

The correction factor of 57.295 converts radians to degrees. Remember that if your data are already expressed as a proportion rather than percentage then you don't need to divide by 100.

MINITAB This assumes that the percentages have been converted to proportions and are in a variable called 'Prop'. From the 'Calc' menu, select 'Calculator…'. In the 'Store results in variable' box, type an appropriate name, such as 'Angular'. Then from the list of 'Functions' highlight 'Degrees' and click 'Select'. Then highlight 'Arcsine' and click 'Select'. 'DEGREES(ASIN(number))' should appear in the 'Expression' box. Then scroll down the 'Functions' list to 'Square root' and click 'Select'. The 'Expression' should now be 'DEGREES(ASIN(SQRT(number)))'. Double click on 'Prop' from the list on the left and the 'Expression' becomes 'DEGREES(ASIN(SQRT('Prop')))'. Click 'OK' to run the transformation.

[If you have the 'commands' enabled ('Editor' menu then 'Enable commands'), and you have already labelled one column as 'Angular', you could type 'Let 'Angular' = DEGREES(ASIN(SQRT('Prop')))' at the MTB> prompt. Or you can input commands using the 'Edit' menu then 'Command Line Editor'.]

Excel Assuming the proportion is in cell A1 the conversion is achieved with the formula '=DEGREES(ASIN(SQRT(A1)))'. DEGREES, ASIN and SQRT can either be typed in directly or selected from 'Paste function' (f_x) under the 'Math&Trig' submenu. The most common error in calculating the angular transformation in Excel comes from the conversion of radians into degrees.

The logit transformation

Logits are needed for logistic regression. The advantage of a logit transformation is that it converts proportional data limited to 0 and 1 to an unlimited scale by using the likelihood of events. The transformation stretches out values that are near 0 and 1. The logit of a proportion, p, is the natural log (ln or $\log_e$) of p/q where q is the proportion that is not p (i.e. $p+q=1$). Note that logits for 0 and 1 are infinite, so will probably give odd results. The logit for 50% should be 0. All values below 50% will be negative, and all above 50% will be positive.

SPSS Assuming the data have been converted to proportions and are in a variable called 'prop', from the 'Transform' menu, select 'Compute Variable'. In the 'Compute Variable' box that appears type a name for the target variable (say, 'logit'). Select either 'All' or 'Arithmetic' from the 'Function group:' list.

Then either, from the list of functions, select 'Ln' and click the up arrow to add it to the 'numeric expression', or type directly into the 'Numeric expression' box. Replace the question mark with 'prop/(1-prop)', replacing 'prop' with the name of your variable. Click 'OK' and the logit-transformed numbers should appear in a new variable.

If converting direct from percentages, the variable to be converted should be divided by 100 as part of the transformation. In this case, assuming your percentage values are in a variable called 'perc', your transformation would be 'LN((perc/100)/(1 – (perc/100)))'.

R This assumes that your data have been converted to proportions and are in a variable called 'prop' and that you want the transformed data in 'logit'. The function for a natural log is 'log()', so the logit for 'prop' is simply:

```
> logit=log(prop/(1-prop))
```

Be warned that values of 0 and 1 will cause the logit to be infinite, giving the error 'Inf'.

MINITAB Assuming the data have been converted to proportions are in column C1, from the 'Calc' menu select 'Calculator…'. In the 'Store results in variable' box, type an appropriate name, such as 'Logit'. From the list of 'Functions' highlight 'Natural log' and click 'Select'. 'LN(number)' will appear in the 'Expression:' box. Replace the text 'number' with 'C1/(1 – C1)', then click 'OK'

[*If you have the* 'commands' *enabled, you could type* 'Let 'Logit'=LOGE(C1/(1 – C1))' *at the MTB> prompt. Or you can input commands using* 'Edit' *menu then* 'Command Line Editor'.]

Excel Assuming the proportion is in cell A2 the conversion is achieved with the formula =LN(A2/(1–A2)). LN can either be typed directly or selected from the 'Insert function' (f_x) under the 'Math&Trig' submenu.

The *t*-distribution

This symmetrical, continuous distribution is related to the normal distribution but is flatter with extended tails. It is the distribution of deviations from the mean divided by the sample standard error of a huge number of samples. As the sample standard error varies between samples the spread is greater than if the deviations were divided by the true standard deviation of the mean (standard error). *t*-distributions have degrees of freedom associated with them that correspond to the size of the sample. So the smallest degrees of freedom of 1 from just two observations will give a very flat distribution and when the degrees of freedom are infininte (i.e. in a sample with an infinite number of observations) the *t*-distribution will recapture the normal distribution.

Confidence intervals

95% confidence intervals (CI) are calculated for samples using t-distributions. (Although when the true σ, standard deviation, is known, or the sample size is huge, the normal distribution can be used.) 95% CI should be preferred to the more usually quoted mean $\pm$ S.E. as the standard error of the mean is only really useful if the sample size is known and then it can be converted to a confidence interval of the required width.

> Be very careful when you see headings such as 'Means and Standard Deviations' as this wording is slightly ambiguous. It usually translates as mean and standard deviation of the observations but is, on occasion, referring to mean and its standard deviation (i.e. standard error).

The chi-square (χ^2) distribution

This is another continuous distribution that is very useful in statistics. Unlike the normal and the t distributions it is asymmetric and varies from 0 to positive infinity. The chi-square distribution is related to variance.

> X^2 is the usual way of expressing sample statistics approximating to χ^2.

The exponential distribution

This is a continuous distribution that is occasionally useful as a null model in biology. It occurs when there is a constant probability of birth, death, increase or decrease. So, for example, if a population of beetles invades a new area they may have an exponential increase in numbers as their rate of increase is constant. As soon as the population stops following the exponential distribution the rate of increase has clearly changed. This may indicate that intraspecific competition has reduced the growth rate or a predator is starting to have an effect.

Exponential distributions can also be used to examine decreasing observations. This is usually called the **negative exponential** distribution. For example, the amount of drug in the bloodstream after an injection may have an exponential decay with 10% being removed every hour. You can test an observed distribution against an expected exponential distribution using a variety of tests of difference.

Non-parametric 'distributions'

It is sometimes better to ignore distributions totally. This is the case when data set is known to be awkward or difficult to transform. The advantages of making no assumptions about the distribution of the data are great as it allows greater flexibility but there are some limitations in the type of statistical tests that can be used and in the power of the tests.

Ranking, quartiles and the interquartile range

In non-parametric tests data sets are usually ranked before they can be examined statistically (computer packages do this for you). If a set of data is put in rank order from the smallest value to the largest then information about the position of the data set or the spread can be gained by inspecting values at certain points in the ranked data set (for example the median is the value of the data point (datum) in the middle of the ranked set).

A quartile is simply the value of the data point that lies a quarter of the way into a data set and it is commonly used to describe the spread of a non-parametric distribution.

The interquartile range is the difference in the values between the data point one-quarter of the way down the ranked list to the point three-quarters of the way down.

Box and whisker plots

Box and whisker plots (also known as, a.k.a., box plots) summarize data where there are no assumptions of distribution. A sample is represented by a box the top and bottom of which represent the upper and lower quartiles (i.e. the box covers the interquartile range). The box is divided at the median value. A line (the whisker) is drawn from the top of the box to the largest value within 1.5 interquartile ranges of the top and the same from the bottom. Any values outside this range are then added as symbols (see Fig. 6.1). These outliers are often identified in some way (they certainly are in the statistical package SPSS) so you can check them. Outliers are the values most likely to have been mistyped!

Descriptive and presentational techniques

6

The techniques in this chapter are presented in roughly the same order as they appear in the key (see Chapter 3).

General advice

Descriptive and presentational techniques serve two rather different purposes. The first is to summarize and display data in the best way possible for a reader to derive information about the data. If this is the intention then the techniques used should be a simple as possible and require the minimum effort from the reader. The second purpose is for researchers to explore their own data. A variety of methods should be employed that show the data from different perspectives. In this way you can become familiar with your data and may be stimulated to pursue new lines of enquiries or test different hypotheses.

This chapter is intended to offer general advice on data presentation and although all the examples are generated in the statistical packages featured in the next chapters there are no detailed descriptions for navigating the menus to generate the figures you see.

Displaying data: summarizing a single variable

Box and whisker plot (box plot)

This is an excellent way of summarizing data, especially if it is not normally distributed. The plot shows the median value as a thick bar, the interquartile range as a box and the full range as the 'whiskers'. Some statistics and graph drawing packages show outliers (data points well outside the range of others) as individuals points. An example is shown in Fig. 6.1.

Choosing and Using Statistics: A Biologist's Guide, 3rd Edition. By Calvin Dytham.
Published 2011 by Blackwell Publishing Ltd.

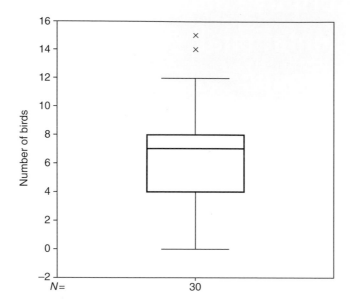

Fig. 6.1 This box and whisker plot was created in SPSS. There were 40 observations of numbers of bird species seen by a single observer from the same point during a fixed time. It shows that the median number was seven and that 50% of the observations were between 4 and 8. Note that there were single observations of 14 and 15 (marked as crosses) and that the axis extends to −2 even though 0 is clearly a lower limit (it is impossible to see fewer than zero species of birds).

Displaying data: showing the distribution of a single variable

It is important that any graphical depiction of data is clear. Usually the easiest and clearest way to display a single set of data is to use a histogram or a bar chart of frequency of occurrence. If you have discrete data then it may be best to display each possible value. However, in most cases it will be necessary to group the data into classes. There is often confusion about the difference between a histogram and a bar chart.

Bar chart: for discrete data

Each possibility is represented on the horizontal axis (abscissa or x-axis), with frequency on the vertical axis (ordinate or y-axis). Gaps between the bars symbolize the discrete nature of the data (see Fig. 6.2 for an example). If there are a very large number of possibilities then a bar chart may be inappropriate as clumping the data into groups will give a better picture of the distribution. If this happens then you have moved to a histogram.

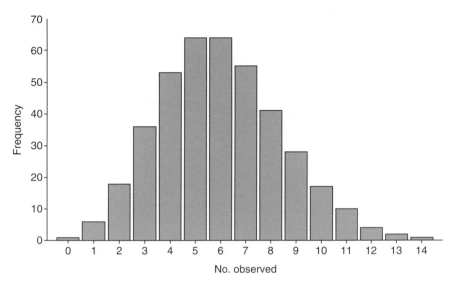

Fig. 6.2 This bar chart is generated using a larger version of the same data set used to create the box and whisker plot in Fig. 6.1. This SPSS chart shows the number of bird species seen in a garden in a 15-min period. There were 400 observations made. Clearly the number of birds can never be below 0 although it might be greater than 15. Observations of this kind will always be integers although there is certainly no requirement for data to be integers to be suitable for bar charts. Gaps between the bars symbolize the discontinuous nature of the data.

Histogram: for continuous data

Observations are grouped into artificial classes. The mid-point of the class is displayed as a label on the x-axis and frequency (number of observations) on the y-axis. No gaps should be left between classes to symbolize the continuous nature of the data. Shading, especially intense shading, should be used sparingly. See Fig. 6.3 for an example.

Number of classes to display in a histogram?

As a rule of thumb use 12–20 classes (categories along the x-axis). However, it is important to employ some common sense. Small samples should rarely need to have 12 classes and huge samples may be grouped into more than 20 classes.

As an alternative rule of thumb use $\sqrt{n}$ classes (where *n* is the number of observations in your sample).

In the example of the beetle elytra above there are 19 classes of 1.5 mm each although only 15 have any observations. This fits with the first rule of thumb. The second rule of thumb suggests 24 classes.

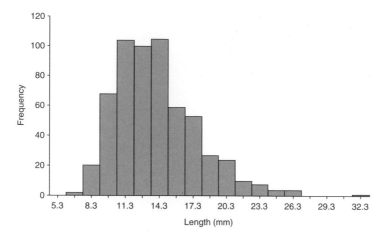

Fig. 6.3 This histogram presents a data set comprising 589 observations of elytron length in a population of beetles. The observations are linear measures and clearly continuous. All measurements were made to the nearest 0.1 mm. Each bar represents a range of values and there are no gaps between the bars. Values on the *x*-axis show the midpoints of the range for half the bars.

Pie chart: for categorical data or attribute data

A pie chart should only be used if the categories have no logical sequence. For example, if the categories are blood groups, species of tree or mutants of *Drosophila* then a pie chart is probably a better method of presentation than a bar chart. However, if the categories have a logical sequence, such as five arbitrarily defined levels of ripeness, then a bar chart will be more informative. An example is shown in Fig. 6.4.

Tip: do not use three-dimensional bars or shadow effects on histograms or bar charts (unless it is for a display and then only in exceptional circumstances). Such effects obscure the data as it is difficult to see exactly where the top of the bar lies. I would also advise against the use of colour unless it is absolutely necessary (although I can't think of an example where I would advocate its use!).

Descriptive statistics

Statistics of location or position

There are several of ways of defining the 'location' of a distribution. It is tempting to focus only on the arithmetic mean as this is the easiest statistic to calculate and the most commonly used. However, it is worth considering some of the alternatives, especially the median.

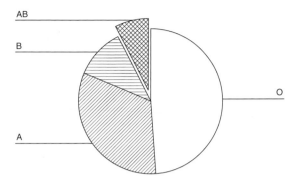

Fig. 6.4 This pie chart, generated in SPSS, shows the blood groups of a sample of 200 people. A pie chart is appropriate for this sort of data because if it was presented as a bar chart the *x*-axis would have no real meaning. Shading is not required but may be used if desired. Slices may be 'exploded' for emphasis as with the AB slice in this example.

Arithmetic mean

This is the 'normal' mean, often called an average and by far the most commonly used measure of location; when written it is usually denoted as *x-bar* (i.e. $\bar{x}$), an estimate of the true mean, which is represented by the Greek letter, μ (mu), sometimes written as $μ_x$.

Geometric mean

This is the antilog of the mean of the logged data; it is always smaller than the arithmetic mean. The most commonly encountered use of this statistic is when data have been logged or when data sets that are known to be right skewed are being compared.

Harmonic mean

This is the reciprocal of the mean of the reciprocals and is always smaller than geometric mean. This type of mean is rarely needed.

Median

This is the middle value of a ranked data set. After the arithmetic mean it is the next most commonly used measure of location. It is the measure highlighted in box and whisker plots. If all the data are put into rank order (arranged in a list in from the largest value to the smallest) the median is the value associated with the middle ranked item (halfway down the list).

Mode

This is the most 'fashionable' value in a set of data; the value that occurs most frequently. It can be used with any type of data, even categorical data.

Variables with one clear mode are said to be unimodal.
Distributions with two peaks are bimodal.
Distributions with more than two peaks are multimodal.
The trough between two modes is sometimes called the antimode (Fig. 6.5).

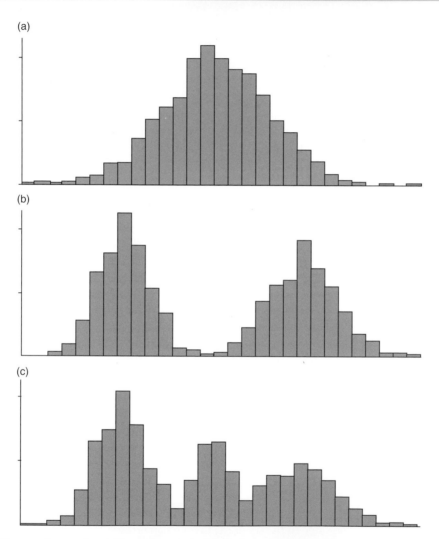

Fig. 6.5 Three rather different frequency distributions. (a) There is a clear single mode of a unimodal distribution. (b) There are two almost totally distinct distributions, giving a bimodal distribution. This might indicate two separate populations, different genders or different species. (c) The pattern of the frequency distribution is even more complex. There are three distinct modes making a multimodal distribution. This may indicate, for example, three cohorts of recruitment into a population.

One of the problems with the use of the mode is that it is rarely suitable if the observations are made with any degree of precision (e.g. femur length to the nearest 0.01 mm) as there will be a much lower chance of an observation being repeated. Therefore the mode should only be used when there are either a very large number of observations or a fairly small number of possible values.

Note: in any unimodal, symmetrical distribution (for example, a perfect normal distribution) the mean, median and mode are all the same (see Fig. 5.3 for what happens in an asymmetric distribution).

Statistics of distribution, dispersion or spread

There are several ways to display the distribution or spread of a set of observations. However, it is important that the measure used is appropriate to the data *and* the statistic of location (e.g. median) used.

Range

This is the most basic measure of dispersion and is simply the difference between the largest and smallest observations in a sample. It is usually quoted as the smallest and largest value (e.g. range = 9.76–15.23 cm).

Interquartile range

This is a non-parametric measure of dispersion that works on the ranked data. It is the difference between the value of the data item (datum) 25% of the way down a ranked list and the one 75% down. These values are called quartiles. The interquartile range is much more useful than the range as it is unaffected by outliers. Unlike many other measures of dispersion, the interquartile range is not necessarily symmetrical about the median. The quartiles are often given the codes 'Q1' and 'Q3'.

Variance

The variance usually refers to the sample variance, s^2, which is an estimate of the true variance, σ^2 (sigma squared). It is the mean of the squared deviations of observations from their arithmetic mean. Variance is rarely used as a descriptive statistic as it is not in the same units as the original observations. However, many statistical tests use variance in their calculations.

Standard deviation (SD)

This is usually an estimate, s, of the true standard deviation, σ (sigma). It is the square root of the variance. This is commonly used as a descriptive statistic as it is in the same units as the original measurements or observations. However, confidence intervals should be used if comparisons of different sets of observations are required.

Standard error (SE)

By convention this is short for 'standard error of the mean' (i.e. the standard deviation of a distribution of means for repeated samples from a population). Standard errors are often quoted with means although this is probably because they are small rather than for any good statistical reason! If several samples are to be compared then the confidence interval should be preferred. If a measure of the variation in the sample is required then standard deviation is better.

> In theory there is no difference in calculation between a standard error and a standard deviation, just that the former measures the standard deviation of a hypothetical sample of means.

Confidence intervals (CI) or confidence limits

These are derived from the standard error of the mean. Confidence intervals are the most useful measure of the dispersion of a distribution.

If a sample from a population is very large then the true mean of the population is 95% likely to lie within 1.96 standard errors of the sample mean. This region is called the 95% confidence interval as you are 95% certain that it contains the true mean of a population.

As samples get smaller then the multiplier used gets larger and the confidence intervals get wider. (If you have statistical tables it is easy to determine the required multiplier as it is derived from the t-distribution.) Confidence intervals are always symmetrical about the arithmetic mean. They are to be preferred over standard errors if several sets of observations are being compared.

Coefficient of variation

This is used to compare the amount of variation in populations with different means where direct comparisons of the standard deviations (s) are difficult to make as they are confounded by differences in scale. The coefficient of variation is usually denoted V or CV. $CV=(100s)/\text{mean}$ and is usually expressed as a percentage.

Other summary statistics

There are other components of shape of the distribution of observations that can be interpreted easily. Knowledge of the skewness of a data set is particularly useful.

Skewness

This is a measure of the symmetry of a data set. If the data set is symmetrical then the value of skewness will be 0. If there is a tail to the right it will be positive; if there is a tail to the left it will be negative. Meaningful values for skewness are only possible if there are more than 30 (and preferably a lot more) observations in the data set. Normal distributions are symmetrical and consequently have a skewness of 0. Skewness is discussed in Chapter 5, page 41.

Kurtosis

This is a measure of the shape of a distribution. It tells you whether there are more observations around the mean or less when compared to a normal distribution. Meaningful values are only possible if there are more than 100 observations in the data set. Kurtosis is discussed in Chapter 5, page 43.

Using the computer packages

General

All statistical packages will give summary statistics for sets of observations. However, generating exactly the set of statistics you are interested in may take several steps. The less frequently used statistics, such as kurtosis, may not be available.

SPSS In this package the data may appear to be in the same spreadsheet form as a package such as Excel but the approach is rather different, as the statistics are not displayed on the spreadsheet but in a separate window. The data should be in a single column with an appropriate label. To change the column label simply double click on the column name ('VAR00001' by default), or click on the 'Variable View' tab, and replace the 'Name' with something more suitable (you are limited to eight characters, spaces are not allowed and you should use the 'Label' column if you need to add a more descriptive name). The screen shot in Fig. 6.6 shows the various measures of dispersion that are available under the 'Descriptives...' options in the 'Descriptive Statistics' submenu of the 'Analyze' menu. The default selections are shown which include the rarely useful minimum and maximum (Fig. 6.6). Once you have chosen the options you want the statistics are displayed in the 'Output' window. As you proceed through an SPSS session output accumulates in this window. This is very useful as you can go back and check results of previous tests very easily.

Further descriptive statistics can be accessed. From the 'Analyze' menu select 'Descriptive Statistics' then 'Frequencies...'. Select the 'Statistics...' button in the dialogue box and an array of options such as quartiles, skewness, kurtosis, variance, mode and median are available. I suggest you uncheck the 'Display

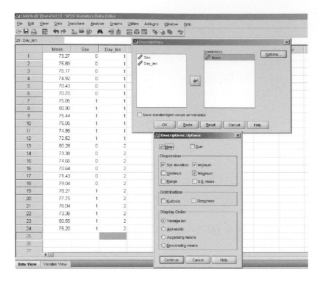

Fig. 6.6 A screen shot from SPSS showing the selection of descriptive statistics available.

frequency tables' unless you want a list of all the values in your variable and how many times they occur.

R Simple descriptive statistics are easy to access in R, although the results are displayed in a rather unhelpful way. In the example below the arithmetic mean of variable 'x' is reported:

```
> mean(x)
[1] 1.703862
```

The functions 'mean()' and 'median()' are obvious. 'var()' gives variance, 'sd()' gives standard deviation, 'length()' give the number of values in a variable and 'sum()' totals the values in the variable, while 'range(x)' will report the lowest and highest value in variable 'x'.

The function 'summary()' gives the mean, median, maximum, minimum and quartiles. Other descriptive statistics are available if packages are installed. Search for 'geometric mean', or 'standard error of the mean' and install the relevant package.

Alternatively you could construct a small script in R to calculate the summary statistic.

The geometric mean is the antilog [the function is called 'exp()' in R] of the mean of the log values of a set of data that can be very easily written in R. The following function calculates the geometric mean of variable 'x':

```
> exp(mean(log(x)))
```

Skewness, or asymmetry, of the data in variable 'x' can be accessed by:

```
> boot::k3.linear(x)
```

There is no function for the harmonic mean, but this can be easily constructed. Assuming the data are in a variable 'x' the harmonic mean is:

```
> 1/(mean(1/x))
```

Simple charts are very easy to access in R. A pie chart can be drawn with the function 'pie()', and the data and labels can be passed to the function or set within the function, for example:

```
> pie(c(12,4,25),labels=c("Ash","Oak","Elm"))
```

The basic plotting function in R is 'plot()' and this is extremely versatile. Again the data can be passed direct to the function, although that would be unusual. A simple demonstration would plot a set of numbers against their squares:

```
> plot(c(1:10),c((1:10)^2))
```

Note that '1:10' gives the numbers 1–10, whereas '(1:10)^2' gives the squares of 1–10. Axis labels can be added using the syntax 'xlab="label text"' within the plot function:

```
> plot(c(1:10),c((1:10)^2), xlab="X axis", ylab="Y
axis")
```

To get help on the options available simply type:

```
> ?plot
```

MINITAB   The data for a single variable should be in one column in the spreadsheet section of the package. The variable should be named appropriately in the cell under 'C1'. Spaces are allowed as part of the label. To get simple descriptive statistics go to the 'Stat' menu and select 'Basic statistics' and then 'Display Descriptive Statistics…'. Move the name of the column with the data from the list on the left into the 'Variables:' box using the 'Select' button. Either click 'OK' now or take a detour to either the 'Graphs…' options first to request some graphical output or to the 'Statistics' options to add output.

Descriptive Statistics: Height

```
Variable        N  N*    Mean  SE Mean  StDev  Minimum      Q1  Median      Q3
Height         24  0   12.985    0.253  1.239   11.000  12.043  12.550  13.750
Variable  Maximum
Height     15.630
```

This is the basic output that is generated:

All the basics are reported here: the number of observations in the data set ('N'), the arithmetic mean ('Mean'), largest and smallest values ('Maximum' and 'Minimum'), standard deviation ('StDev') and standard error ('SE Mean') as well as the 'non-parametric' distribution statistics: median and the upper and lower quartiles ('Q3' and 'Q1'). Quartiles are explained further elsewhere in this chapter. One option in the 'Statistics' options is the rather unusual trimmed mean 'TrMean' where the tails of the distribution, top and bottom 5% in this case, are removed before a mean is calculated. This makes the estimate of the mean less likely to be affected by outliers.

If the 'Graphical summary' is selected as an option in the 'Graphs…' options box, there is considerably more output to assess (shown in Fig. 6.7).

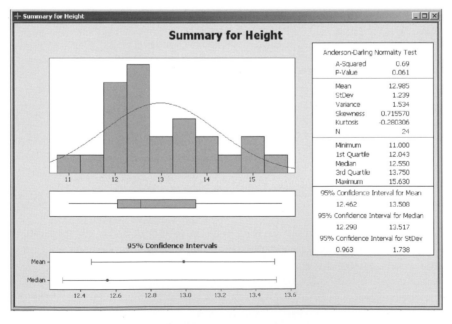

Fig. 6.7 A screen shot from MINITAB when the 'Graphical summary' option has been chosen to display descriptive statistics.

This output contains much of the same information as the non-graphical version but with some extras. In the mass of output on the right of the output the first thing is the 'Anderson-Darling Normality Test'. This is a test to determine whether the data in question deviate from a normal distribution. The 'A-Squared' value is the output from a test and 'P-Value' is the probability of seeing data that, or more, extreme if they are a sample from a normal distribution. If the *P*-value is less than 0.05 then this means it is unlikely to be normally distributed and therefore parametric statistics should not be used.

After this test comes the more usual descriptive statistics of arithmetic mean, standard deviation, variance and then the measures of the shape of the distribution – skewness and kurtosis – and the number of observations, 'N'. Next comes some information about the data arranged in rank order. The value of the smallest and largest observations and then observations one-quarter (first quartile), half (median) and three-quarters (third quartile) of the way down a ranked data list.

The last section gives 95% confidence intervals for three of the descriptive statistics: 'Mean' (the arithmetic mean), the median and 'StDev' (the standard deviation). On the left of the output are three graphs. First is a histogram of the raw data with a normal distribution superimposed on it (the normal distribution shown has the same mean and standard deviation as the data). Then comes a box and whisker plot of the data (described elsewhere in this chapter) using the same scale as the histogram and finally graphical representations of the mean and median with their 95% confidence intervals.

Excel In this package you have to assign a cell of the spreadsheet to contain the summary statistic you require. Assign a cell by clicking on any empty cell. Then you identify the cells that contain the variable (raw data) that you are interested in and the statistic appears.

For example: your data, containing 100 observations, has been typed into the first column of the spreadsheet (column A). The first cell has the title of the variable and the actual observations are in rows 2 to 101. To calculate the arithmetic mean of this variable you go to any cell and declare its contents as '=AVERAGE(A2:A101)'. (As you can see Excel calls the arithmetic mean the 'average'.) The median can be calculated as '=MEDIAN(A2:A101)', the geometric mean by '=GEOMEAN(A2:A101)' and the harmonic mean as '=HARMEAN(A2:A101)'.

These and other summary statistics are easily accessed using the 'Insert function' (f_x) facility of the package, mostly in the 'Statistical' submenu. Or, once you have learned a few of the function codes you could just type them in. For instance '=STDEV(A2:A101)' to get the standard deviation reported in an empty cell. Most of the summary statistics mentioned in this chapter are readily available in Excel, including skewness (SKEW), kurtosis (KURT) and mode (MODE). Confidence intervals require more work, but are possible using CONFIDENCE. The command '=CONFIDENCE(0.05,STDEV(A2:A101),100)' will give the

95% confidence interval of the 100 items of data in column A, the parameters 0.05 and 100 set one minus the size of the confidence interval and the sample size respectively.

The interquartile range is not given directly in Excel, but it is easily calculated. '=QUARTILE(A2:A101,1)' will give the value of the first quartile (i.e. the datum 25% up the data set when sorted) and '=QUARTILE(A2:A101,3)' will give the third quartile (75%). The interquartile range is the difference between these two values, so in Excel that would be '=QUARTILE(A2:A101,3) – QUARTILE(A2:A101,1)'.

Displaying data: summarizing two or more variables

Box and whisker plots (box plots)

These are a good way of comparing two variables. They allow direct visual comparison of both the location and the dispersion of the data. An example of the use of two box plots is shown in Fig. 6.8.

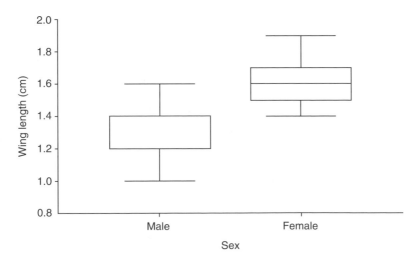

Fig. 6.8 In this SPSS-generated figure the sample of observations of wing lengths of a moth are divided into two groups by gender. As in most insects, the females are considerably larger than the males and although there is some overlap in the whiskers there is no overlap of the interquartile range of the two groups. Note that for the males the median and lower quartile are superimposed, showing that 25% of the observations for males were almost of the same value.

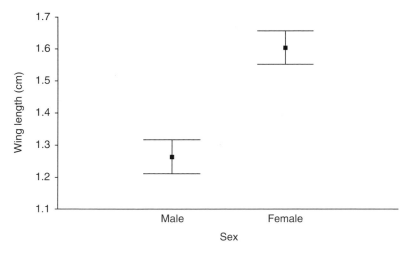

Fig. 6.9 This figure, also generated in SPSS, uses the same moth wing lengths as Fig. 6.8. The means for males and females are represented by filled squares and the whiskers are error bars that extend to cover the 95% confidence interval for the mean (i.e. there is a 95% chance that the true mean of the population lies between the extremes shown). There is no overlap between the whiskers, suggesting that the groups are likely to be highly significantly different.

Error bars and confidence intervals

A similar way of looking at the same data is to display the arithmetic mean and some measure of the dispersion of the data. An example of the use of mean and confidence interval is given in Fig. 6.9. Note that the interquartile range is not symmetrical about the median (Fig. 6.8) whereas the 95% confidence intervals (or standard deviation if we had chosen to display that instead) are symmetrical about the mean (Fig. 6.9).

You can display more than two groups using these methods. They provide a very powerful method of showing differences and similarities between many groups. In the example here there is almost no need for any further statistics, as the difference between males and females is so striking!

Displaying data: comparing two variables

Associations

If two observations are made from a single individual (e.g. the 'individual' is a stream, and the water pH and stream flow have been recorded), before any statistics are applied it is best to get a 'feel' for the observations by a graphical representation of the data.

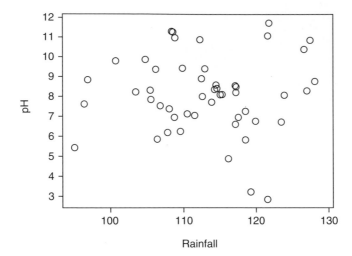

Fig. 6.10 The scatterplot of pH and rainfall from a range of sites shown here has been created in MINITAB using the default options. Clearly there is no obvious relationship between the two variables.

Scatterplots

The simplest way to display a relationship between two variables is to use a plain scatterplot (Fig. 6.10). This assumes that two observations on the same row in the package are two measurements from the same 'individual'. An 'individual' can be almost anything: sampling station, greenhouse, pair or single bone.

It is important that all figures should have appropriate axis labels on them. They should also be accompanied by a figure legend that makes the plot interpretable without reading the relevant section of the text.

Do not add extra information that is not relevant or appropriate. For example, many packages offer best-fit lines as a simple option. Do not use these unless (1) you believe there is a 'cause-and-effect' relationship between the variables, (2) you have used regression and you want a graphical accompaniment, (3) you intend to use regression and (4) you wish to use one variable to predict the other.

Multiple scatterplots

A good way to compare observations from two sites where the same variables have been recorded is to use a multiple scatterplot. The axes will be exactly the same as for the single scatterplot but each group will be displayed using a different symbol.

This technique works particularly well for two or three groups and less well for more than that. Choose symbols carefully to allow the groups to be easily

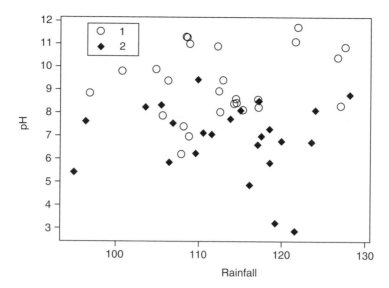

Fig. 6.11 In the example shown here two sets of observations from different study areas are identified with different symbols. A quick glance shows that group 1 is associated with a higher pH than group 2 but there is no obvious difference between the groups on the 'rainfall' axis. An analysis of variance or t-test could be used to determine the statistical probabilities, but the results would only confirm what is obvious from the scatterplot.

distinguishable and make sure that the figure caption makes it clear which symbol matches which group (Fig. 6.11).

More sophisticated use of symbols can convey a great deal of information about several factors on the same scatterplot. For example if the data for two morphological variables are collected and the individuals are divided into groups by sex and species then all this information can be incorporated in a single plot. This can be achieved by using shaded and non-shaded symbols for the two sexes and different shapes for the two species (see Fig. 9.1 for an example).

Trends, predictions and time series

Lines

These should only be used to join points if there is a reasonable assumption that observations could be made between the points (Fig. 6.12). This is perfectly reasonable if the x-axis is temperature with readings made at 15–35°C in steps of 5°C as intermediate temperatures are valid. However, if the x-axis is number of eggs in a nest and the y-axis egg weight then it is perhaps unwise to draw a line linking mean weight at four eggs with mean weight at five as the line will pass though impossible points.

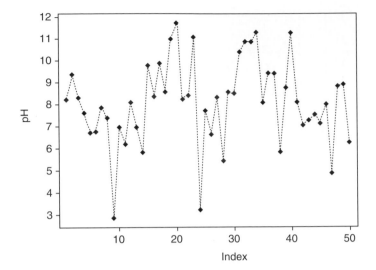

Fig. 6.12 This MINITAB-generated example of a line graphs shows a set of 50 readings of pH made through time at a chemical plant. The time gaps were equal and the observer thought it valid to join the reading made with lines, as there is a reasonable expectation that the intervening times would have intermediate levels of pH.

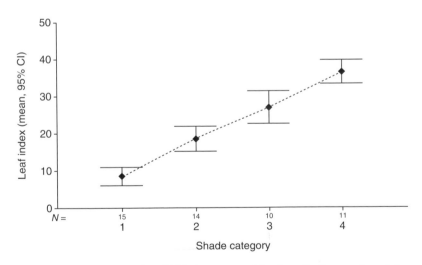

Fig. 6.13 This figure, generated in SPSS, shows a combination of a line graph and the mean and confidence interval approach of Fig. 6.9. In this case there are four levels of the variable shade that can be said to form a valid sequence from light to dark. Observations of leaf shape (a continuous variable) were taken at each of the four shade categories and the mean and 95% confidence intervals are plotted here with the means joined to emphasize the clear trend.

If there are several observations for each point on the x-axis then it is usually better to plot the mean or median with a measure of the dispersion at each point rather than use a scatter of points (Fig. 6.13). The same guidelines for joining means apply as for joining single observations.

It is very easy to deceive a reader by altering scales. For example, if there is a slight but steady increase in the concentration of nitrate in a lake over time then this can be made to look like a rapid increase if the scale on the y-axis starts not from zero but from a value just below the lowest observed value. This kind of manipulation of the reader will work particularly well if there is no measure of the variation given.

Fitted lines

The best way to draw the reader's eye to a relationship between two variables is to use a fitted line of some sort. Indeed, an observer can sometimes be fooled into seeing a relationship in a scatterplot when there is none (Fig. 6.14). For this reason the use of fitted lines should be restricted to circumstances when the line is meaningful. The most common use is to illustrate a relationship between two variables that has been investigated using regression.

One technique is to plot the scattered observations along with the fitted line and then give more information about the regression in the text or the figure legend.

Confidence intervals

These should always be used to show the reliability of a mean value, as shown in Figs 6.9 and 6.13. If a regression line has been calculated then it should

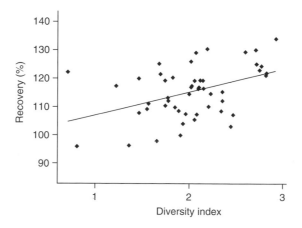

Fig. 6.14 This MINITAB-generated scatterplot of the recovery of biomass after an extreme event against the diversity index before the event shows a slight but non-significant trend. However, the addition of the trend line draws the eye and emphasizes the slight trend, convincing the reader that it is real.

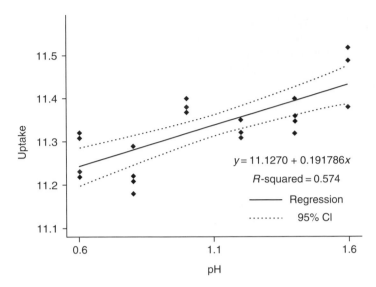

Fig. 6.15 Here the raw data, regression line and the 95% confidence intervals of the regression line are all shown along with some of the regression output from MINITAB. The variable 'uptake' measures the amount of drug passed across the stomach lining of a rabbit at various experimental pH levels. There is a clear relationship: the regression line slope is significantly different from zero and it explains 57.4% of the variation in the uptake observations. The 95% confidence intervals are quite close to the best-fit line confirming that the relationship is robust. Note that the line, quite properly, does not extend beyond the data as predictions can only be safely made within the range of the data.

always be displayed with its confidence intervals. This shows the range within which the line is 95% likely to lie (Fig. 6.15). If the confidence intervals are wide apart then the line is obviously less reliable.

Displaying data: comparing more than two variables

Associations

It may be tempting to use the full capacity of the graphics on the package you are using but there is little or nothing to be gained by plotting a multidimensional plot that is impressive to look at but impossible to interpret.

Three-dimensional scatterplots

This type of figure looks impressive but is quite difficult to interpret for several reasons associated with representing three dimensions in a two-dimensional

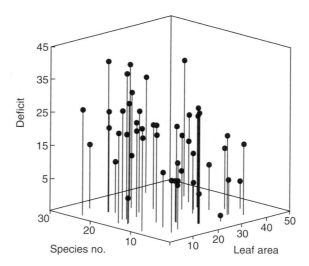

Fig. 6.16 Most statistical packages have three-dimensional plotting ability (this one was plotted using SPSS). The three-dimensional scatterplot is very difficult to interpret even when the relationship is quite strong, and impossible when the relationship is weak. The spikes make the figure very cluttered but are vital to place the point accurately.

medium. First, if there are too many points plotted then those nearest the 'front' will obscure those as the back. Second, as the display medium is two-dimensional all the points need to have 'spikes' to anchor them to the $x=0$, $z=0$ plane. Without these spikes then a point near to the front but high on the y-axis will look *identical* to one near the back but low on the y-axis. An alternative method of spiking is to attach all points to the origin. This occasionally is useful, but normally generates a figure that looks like a bunch of flowers. Finally, there is often no forced perspective, making the arrangement of the axes seem odd. Furthermore points at the back are usually the same size as those at the front and this fools the eye (Fig. 6.16).

Multiple trends, time series and predictions

Multiple fitted lines

Further information may be conveyed if two lines are fitted on the same graph. The advantage of this approach is that lines may be compared directly but the disadvantage is that the message may become confused. I would advise against a tactic I have seen used increasingly which is to have different y-axes for the same x-axis so that the two lines being compared fit sensibly. There are two problems with this type of graph. First it makes the reader see relationships that are not really there and second it is often difficult to see which scale applies to which line.

Surfaces

Many statistical packages include the option to have spectacular three-dimensional surface plots. I would advise against the use of these in almost all situations. The problems of all 'three-dimensional' graphs on two-dimensional surfaces apply with the additional problem that the solid or apparently solid surface totally obscures much of the surface.

The way the points are connected to form the surface is questionable too. For example, if one or two of the axes have data that are normally distributed this means that these observations are contributing a great deal of information in the centre of their range and rather less at the extremes. The surface plot does not reflect this in most cases (except it often betrays this by tending to have smoother edges where the surface is extrapolated from fewer points). Therefore the edges of the surface can be influenced by the extreme points, the very points likely to be measured with less accuracy.

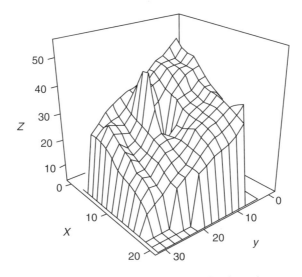

Fig. 6.17 Possibly even more difficult to interpret than the three-dimensional scatterplot shown in Fig. 6.16 is the three-dimensional surface plot, such as this one drawn in MINITAB. This sort of figure can only be interpreted if relationships are very strong or the smoothing algorithm is so strong that all the variation is wiped out of the data. Particular problems of three-dimensional surfaces are that the edges tend to be extrapolated from far fewer data points than the middle and that peaks can obscure a lot of the surface behind them.

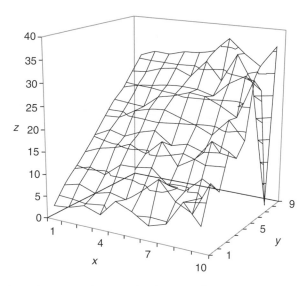

Fig. 6.18 This three-dimensional scatterplot from Excel shows a common problem with interpolation of points. The surface is constructed from an array of 100 values arranged in a 10×10 grid. Unfortunately one value is missing at $x=10$, $y=9$ and this value has been interpreted as having a value of zero.

Remember that a surface, like a joined-up line, can only be used if both the 'x' and 'y' (or 'z') observations can reasonably be expected to have possible intermediate values (Fig. 6.17). A problem with the interpolation of missing data points is shown in Fig. 6.18.

7

The tests 1: tests to look at differences

The tests in this chapter are presented in roughly the order they appear in the key (see Chapter 3).

Do frequency distributions differ?

Questions

There are two basic types of question that can be asked.
1 Does one observed set of frequencies differ from another?
2 Do the observed frequencies conform to a standard distribution?
In the first case the test becomes an analogue of a two-sample test of difference, such as the *t*-test. In the second it is a way of testing observations against expected frequencies, such as in plant-breeding crosses when particular ratios of phenotypes are expected or to test whether organisms are occurring at random by testing against the Poisson distribution. The **G-test**, **chi-square goodness of fit**, **Kolmogorov–Smirnov**, **Shapiro–Wilk** and **Anderson–Darling** tests are the most commonly employed tests to answer these questions and are described below.

G-test

In situations where you have observed frequencies of various categories and expected proportions for those categories that were not derived from the data themselves then the G-test should be the preferred statistic to use. However, it is not many years since this test was shown to be superior to the traditional chi-square goodness of fit approach on theoretical grounds. Consequently it is not supported by any of the packages considered in this book.

If your package supports the G-test then use it and its associated correction factor, the Williams' correction.

In the G-test the ratio of the observed and expected frequencies is calculated. The natural log (ln or $\log_e$) of this ratio is calculated and these values are

Choosing and Using Statistics: A Biologist's Guide, 3rd Edition. By Calvin Dytham.
Published 2011 by Blackwell Publishing Ltd.

multiplied by the number observed, summed, then doubled. This value of G is then compared to a chi-square distribution with one fewer degrees of freedom than the number of categories.

An example

A dihybrid cross of sweet peas has four categories of plant types with an expected ratio of 9:3:3:1. This ratio was not generated by the data so when the data are collected it should be compared to the expected values using the G-test. Two hundred plants were collected. The number of plants in each of the four categories is given below.

Tall and pink	Tall and white	Dwarf and pink	Dwarf and white
108	35	46	11

SPSS This test is not available in SPSS.

R This is quite a simple test and can be achieved in a few steps in R.
1 Enter the data into a variable, here called 'obs':

```
> obs<-(c(108,35,46,11))
```

2 Then enter the expected frequencies into a variable, here called 'expected':

```
>expected<-(c(9,3,3,1))
```

3 These expected frequencies need to be converted to expected frequencies accounting for the size of the sample, so each value in 'expected' is divided by the total of the values in 'expected' and this is multiplied by the total number of observations:

```
>expected_freq=(expected/sum(expected)*sum(obs))
```

4 This is then converted to a log ratio stored in a variable called 'lnratio':

```
>lnratio=log(obs/expected_freq)*sum(obs)
```

5 To check all is well the values held in 'lnratio' can be displayed:

```
> lnratio
[1] -4.408775 -2.414751 9.397821 -1.406167
```

6 The result of the test is the absolute value (i.e. it must be positive) of double the sum of the values in 'lnratio'; here I've called it 'g':

```
>g=2*abs(sum(lnratio))
```

7 This value is then compared to a chi-square distribution with degrees of freedom one fewer than the number of categories. In R the function 'pchisq()' is used for this. This function requires two values: the value to be compared and the degrees of freedom. The result is the *P*-value:

```
>1-pchisq(g,3)
```

Remember that these instructions only apply when there are four values and you should adjust to accommodate different numbers of categories.

MINITAB This test is not available in MINITAB.

Excel As this is a relatively simple calculation it is ideal for a spreadsheet like Excel. The most likely source of error when carrying out a G-test is in the calculation of the expected frequencies. The total of the expected frequencies should be exactly the same as the number observed.

1 Label five columns as: observed, expected ratios, expected frequencies, ratio and ln ratio.

2 Enter the four observed values in cells A2–A5. (Your data may have more than four categories and therefore all references to row five here should be adjusted to your data set.)

3 Enter the values 9, 3, 3 and 1 in cells B2–B5.

4 In cell A7 use '=sum(A2:A5)' to sum the number of observations.

5 Copy cell A7 across to B7: '=sum(B2:B5)'.

6 In cell C2 the expected frequency of tall and pink plants is needed. This should be 9/16 of the total number of observations. Use '=B2/B$7*$A$7', which will give 112.5 in this example, and copy this cell down to C5. Note that the '$' are important as they hold the row or column or both in place when copying. 'F4' can be used to cycle through $ options.

7 Calculate the ratios in D2 as '=A2/C2'. Copy this cell down to cell D5.

8 Calculate the natural logs and multiply by the number of observations (A2) in E2 as '=A2*ln(D2)'. Copy down to cell E5.

9 Sum the values in column E in cell E7 and double it: '=2*sum(E2:E5)'. The example should give a value of 2.336. If your data gives a negative value adjust the contents of cell E7 to '=−2*sum(E2:E5)' and the value should become positive.

10 The value in E7 should be looked up on a chi-square table. Excel does this with '=CHIDIST(E7,3)' (or select CHIDIST from the 'Paste function', Statistical submenu). In this example there are three degrees of freedom. The result is

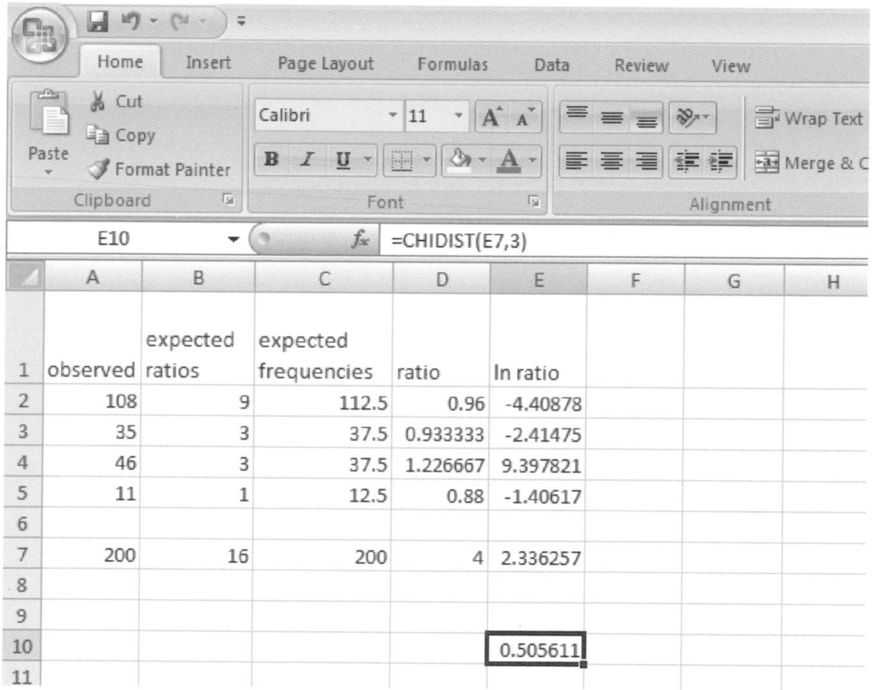

	A	B	C	D	E	F	G	H
1	observed	expected ratios	expected frequencies	ratio	ln ratio			
2	108	9	112.5	0.96	-4.40878			
3	35	3	37.5	0.933333	-2.41475			
4	46	3	37.5	1.226667	9.397821			
5	11	1	12.5	0.88	-1.40617			
6								
7	200	16	200	4	2.336257			
8								
9								
10					0.505611			
11								

Fig. 7.1 Calculating the value of G using Excel.

$P = 0.505611$, so the null hypothesis is not rejected (see Fig. 7.1). Your calculation should have one degree of freedom fewer than the number of categories.

11 Save your Excel sheet as it will be easy to recycle for future G-tests by altering the values in columns A and B (where necessary). If more categories are required remember to adjust all the summing steps to cover the required ranges and adjust the degrees of freedom.

Chi-square test (χ^2)

Often known as the chi-square goodness of fit, this test is one of the most widely used in the whole of biology. It is also the statistical test you are most likely to be familiar with. You will usually present the data in a table showing the observed and expected frequencies for various categories. These categories can be single outcomes or groups of possible outcomes. It is customary to use grouping of categories to ensure that none of the expected values is less than 1 (some authors, erring on the side of caution, suggest 5). The expected values can be derived from a distribution such as the Poisson or negative binomial, they can assume that all categories are equally likely (a flat or rectangular distribution), they can be derived from a specific null hypothesis of ratios or they can be derived from another set of data. In all cases the null hypothesis (H_0) will be that the observed and expected frequencies are not different from each other. The chi-square test may also be used as a test of association (see Chapter 8).

An example

A very common starting point for investigations in biology is to determine whether events or observations are occurring at random. If events are truly random then they should follow a Poisson distribution (see Chapter 5 for more details). The chi-square test allows you to compare observed data with the expected data if following a Poisson distribution with the same mean. In this case the number of lice found on adult char is recorded. All observations were taken from a single catch of 98 fish. If the lice attach themselves to the fish at random then they will follow a Poisson distribution and the chi-square will not be significant (i.e. supporting the null hypothesis that lice attack fish at random). If the result is significant this indicates that lice do not attack randomly and a new hypothesis should be formulated.

Note that in this instance we are comparing a set of observed frequencies against a set of expected frequencies derived from a Poisson distribution that has the same mean as the observations. Therefore the expected frequencies are not independent of the observed and this loses us a degree of freedom.

No. lice/fish	0	1	2	3	4	5	6	7	8+
No. observations	37	32	16	9	2	0	1	1	0

SPSS First I should point out that processing this type of data for a goodness of fit chi-square in SPSS is not easy unless you wish to fit the distribution to a uniform one (i.e. all categories are expected to have the same number of observations). If you have this sort of data use the Kolmogorov–Smirnov test to answer the question in SPSS. However, I will go through the procedure anyway assuming you don't know how to calculate Poisson 'expecteds' by hand (perhaps the faint hearted should move on to the Kolmogorov–Smirnov test now!).

1 Make sure that the data are in a single column of the actual data, not the frequencies. In this case there will be 98 rows in the data set; one for each fish. Label this column 'no_lice'.

2 Determine the mean number of lice per fish using the 'Analyze' menu, then 'Descriptive Statistics ...' and selecting either 'Descriptives...' or 'Frequencies...' before moving 'no_lice' into the 'Variable(s)' box. The output will confirm the mean as 1.14. The next steps show how to calculate the expected frequencies in SPSS. If you can do this by hand or in a spreadsheet I suggest you do so and skip directly to step 6.

3 Generate a new column with all the expected frequencies in it (e.g. 0, 1, ... 7) in separate rows. Label this column 'freq'.

4 Use the 'Compute Variable…' which is under the 'Transform' menu to bring up a dialogue box. In the 'Target variable' box enter, say 'exp1'. Select 'All' or 'PDF & Noncentral PDF' in the 'Functions group:' box. Then scroll down the list of functions until you reach 'Pdf.Poisson'. (*Note*: PDF stands for probability density function.) Select this and move it into the 'Numeric Expression:' box with the up arrow. Replace the second question mark with the mean you calculated in step 2 (i.e. 1.14 in the example). The first question mark ('q') should be replaced with the name of the variable you created in step 3 (i.e. 'freq'). The expression should be 'PDF.POISSON(freq,1.14)'. Click 'OK'. A new variable will appear on the spreadsheet with numbers starting from 0.32. This first number is the probability of getting a zero in a Poisson distribution with a mean of 1.14.

5 One more calculation step is required. Go to the 'Transform' menu and select 'Compute Variable…' again. Insert a new label in the 'Target variable', say 'expected' as this is going to be the true expected value. Select 'exp1' from the list on the left and move it to the 'Numeric expression' box. Then add '*98' (or whatever your sample size is if you are not using the example) outside the parentheses. This will multiply the values by the total number of observations, turning your expected frequencies into expected numbers of lice.

6 You will notice that all expected frequencies for more than four lice per fish are less than one and should be grouped together to form a category of 'four and above'. Write down the expected frequencies for the five categories.

No. lice per fish	0	1	2	3	4+
No. observations	37	32	16	9	4
Expected no. observations	31.3	35.4	20.4	7.8	2.8

7 As all values above three lice per fish have been grouped for expected frequencies, this must now be done for the actual data. In SPSS you can either do it by hand or use the 'Recode' option under the 'Transform' menu. Selecting either 'Recode into same variables…' (over-writing the original data) or 'Recode into different variable…'. If you choose the latter, move 'no_lice' into the 'Numeric variable → output variable' box. Then type a name for your new variable in the 'Output variable, Name:' box and click 'Change'. Your new name appears after the old name in the main box. Then click 'Old & New Values…'. This brings up a bewildering set of options that are actually extremely useful in SPSS. In the 'Old value' section on the left select 'Range, value through highest'. Put a 4 in the box. Then in the 'New value' section on the right put a 4 in the 'value' box and click 'add'. This will put '4 thru Highest → 4' in the 'old → new' section (meaning that all 4s or higher will become 4s).

Finally click 'All other values' on the left, 'copy old values' on the right, and 'add' to keep the rest as they were. 'ELSE → Copy' appears in the 'old → new' box. Click 'Continue' here and then 'OK' in the next window to create the new variable.

8 Finally, choose 'Analyze', 'Nonparametric tests' and 'Chi-Square...' to bring up the 'Chi-Square Test' window. Move the variable you created in step 7 into the 'Test variable list:' box. Unfortunately you have to enter the expected frequencies one at a time. Click on 'Values:' in the 'Expected Values' area and enter the expected frequencies, starting at the one for zero (i.e. 31.3 in the example) and clicking 'add' after each one before clicking 'OK' to run the test. This is the output you should expect.

Frequencies

Chi-Square Test

no_lice

	Observed N	Expected N	Residual
.00	37	31.4	5.6
1.00	32	35.5	–3.5
2.00	16	20.5	–4.5
3.00	9	7.8	1.2
4.00	4	2.8	1.2
Total	98		

Test Statistics

	no_lice
Chi-Square	3.002[a]
df	4
Asymp. Sig.	.557

a. 1 cells (20.0%) have expected frequencies less than 5. The minimum expected cell frequency is 2.8.

The data have been reduced to just five categories and each has an associated expected frequency. The residual is the difference between observed and expected. At the bottom is the actual chi-square statistic, the associated degrees of freedom (4 in this case as, once the categories had been clumped, there were five categories) and finally the significance value (P-value).

> *Note*: as the mean of the distribution we are comparing to our data was calculated from the sample we should lose a further degree of freedom to give a total degrees of freedom of 3. The package does not account for this as it is calculated before the test is applied and therefore SPSS 'knows' nothing about it.

Despite the fact that there are more high values and zero values than you would expect (indicative of a clumped distribution) the probability is well above 0.05. This indicates that the deviation from the Poisson expectations is non-significant and therefore we have no reason to reject the null hypothesis that lice attack fish at random in this population.

R As with SPSS and MINITAB the default chi-square test in R, 'chisq. test()', assumes that the number of observations per category is equal. There are several ways to tell it otherwise, and here I'm going to avoid using a predefined chi-square function as it gives more control over the amalgamation of categories. To generate the expected values for the table we will make use of the function 'dpois()' that gives the proportion of observations with a given value in a Poisson distribution with a given mean. For example, >dpois(1,2.2) will return 0.2437699, which is the proportion of values that will be 1 in a Poisson distribution with a mean of 2.2.

1 First make a variable that holds the integers from zero to eight representing the number of lice: >number<-c(0:8)

2 Then input the numbers of observations. Perhaps these will be held in a text file and imported into R, but in this case I'll assume that the values from the tally table need to be input to R manually: >obs<-c(37,32,16,9,2,0,1,1,0)

3 We need to know the total number of observations and the total number of lice to calculate the mean number of lice per fish. I'll do this in one go in R, then confirm the correct value:

```
>meanlice=sum(obs*number)/sum(obs)
>meanlice
[1] 1.142857
```

4 Now we know the mean number of lice per fish we can calculate the expected proportion of fish that will have zero, one, two, etc. lice using 'dpois()', which assumes that lice attack fish at random leading to a Poisson distribution. As what is required is an expected number of fish, the proportions are multiplied by the number of observations:

```
>expected_freq=sum(obs)*dpois(number,meanlice)
>expected
```

```
[1]  31.2526 35.7175 20.4100 7.7752 2.2214 0.5078 0.0967
[8]  0.0158 0.0023
```

5 In this example there are several categories that have very low values. The advice given for chi-square tests is that no expected values should be less than 1 and no more than a quarter of the expected values should be less than 5. This means only the first four or five categories should be used in this case. It is important that the sum of the values in the observed and expected variables are the same, so care should be taken when working out the value in the 'four and above' category. Here we know that the total number of observations is 98, so the value in 'four and above' should be the sum of the values in categories 0 to 3 taken away from 98, or 'sum(obs)'.

```
>sum(obs)-sum(expected_freq[1:4])
[1]  2.8444
```

This can be inserted as the fifth element in the 'expected_freq' variable:

```
> expected_freq[5]=sum(obs)-sum(expected_freq[1:4])
```

6 Now we have five categories of expected values, 0 to 3 and '4 and above'. This needs to be reflected in the observed data as well. So the value of the fifth element of the obs variable should be the sum of all observations of four and above. Or, it could be done by subtraction from the total number of observations:

```
>obs[5]=sum(obs)-sum(obs[1:4])
```

This means that the value of the '4 and above' category is now 4.
7 To calculate the value of chi-square the simple formula of the sum of (observed–expected) squared over expected is implemented for the first five elements of these variables.

```
>v=(obs-expected_freq)^2/expected_freq
>chisquare=sum(v[1:5])
>chisquare
[1]  3.059116
```

8 Finally we need the *P*-value associated with a chi-square value of 3.059116 with three degrees of freedom. The degrees of freedom is three rather than four because there has been some clumping of categories. 'pchisq()' returns the cumulative probability of getting a chi-square value of that or lower, so to calculate the *P*-value we need '1-pchisq()':

```
>1-pchisq(chisquare,3)
[1] 0.3862
```

Here the *P*-value is well above 0.05 so we don't reject the null hypothesis that the frequencies of number of lice per fish follows a Poisson distribution.

MINITAB Calculating a goodness of fit to a Poisson distribution is surprisingly awkward in MINITAB, although not quite as awkward as in SPSS. However, it is a test that you will wish to carry out in many circumstances.

1 Make sure that the data are in a single column of the actual data, not the frequencies. In this case there will be 98 rows in the data set: one for each fish. Label the column 'No lice'.

2 Determine the mean number of lice per fish using the 'Stat' menu, then 'Basic statistics', then 'Display Descriptive statistics'. Move 'No lice' into the 'Variables' box, either by double clicking on the variable name or using the 'Select' button. The output will confirm that there are 98 observations and give the mean as 1.143.

(Or, if the command interface is enabled, type 'Describe 'No lice'' *or* 'Describe C1' *at the MTB> prompt. Or you can input commands using* 'Edit' *menu then* 'Command Line Editor'.)

The next steps show how to calculate the expected frequencies for a Poisson distribution in MINITAB. If you know how to do this by hand or in a spreadsheet you can skip directly to step 6.

3 First you should generate a table of the tallied observations: Go to the 'Stat' menu, then 'Tables' then 'Tally individual variables...'. Move 'No lice' into the 'Variables' box and make sure that the 'Counts' box is checked. Click 'OK'. This output will appear in the 'Session' window:

Tally for Discrete Variables: No lice

No lice	Count
0	37
1	32
2	16
3	9
4	2
6	1
7	1
N=	98

Note that as there were no fish with five lice, there is no count for five in the table. Now you can either cut and paste the two columns of figures from the 'Session' window into columns C2 and C3 of your MINITAB spreadsheet (remembering to click on the 'Use spaces as delimiters' option) or you can type the numbers in directly.

(Or type 'Tally c1;' *at the MTB> prompt, followed by* 'Store c2 c3.' *at the SUBC> prompt. Remember to type a semicolon at the end of MTB> command if you want to bring up the SUBC> prompt.)*

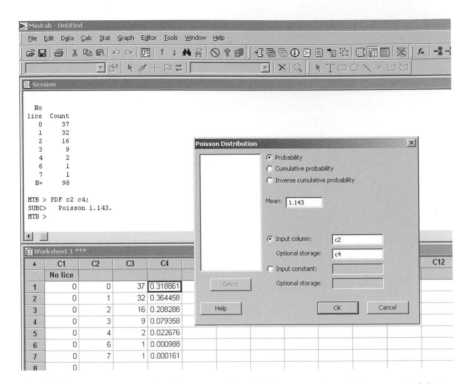

Fig. 7.2 Generating expected values using the 'Tally' command and Poisson probabilities in MINITAB.

4 Now we need to generate expected numbers of lice for a Poisson distribution with the same mean as the sample. Go to the 'Calc' menu, then 'Probability distributions' and then 'Poisson…'. In the dialogue box type the sample mean in the 'Mean:' box, type 'c2' in the 'Input column:' box and 'c4' in the 'Optional storage:' box. If you don't select a storage column the output only goes to the 'Session' window (Fig. 7.2). Click 'OK'.

(Or type 'PDF c2 c4;' *at the MTB>prompt, followed by* 'Poisson 1.143.' *at the SUBC>prompt. Replace 1.143 with the mean of your sample. Or you can input commands using* 'Edit' *menu then* 'Command Line Editor'.)

5 To convert the probabilities generated into numbers you need to multiply by the total number of observations in the sample (98 in this example). Go to the 'Calc' menu, then 'Calculator…'. In the dialogue box type c5 in the 'Store result in variable:' box. Then type 'c4 * 98' in the 'Expression:' box (replacing 98 with the number of observations in your sample). Click 'OK'.

(Or type 'Let c5 = c4 * 98' *at the MTB>prompt.)*

6 Chi-square tests should not have expected frequencies that are less than one. In the example the expected frequencies for 6 and 7 are both less than one. Also, because there were no fish with five lice in the example there is no expected frequency for 5 in the column. In this example the best strategy is to pool all observations of 4 or more into a single observed and expected value. You can do this in several ways, either on a calculator, by hand or by using the 'Calculator...' from the 'Calc' menu, selecting a row and typing the sum required. In this example selecting row 5 and typing '98–31.248–35.717–20.412–7.777' in the 'Expression box'. The numbers are the expected frequencies of 0, 1, 2 and 3 lice per fish respectively. The result should be 2.85, being the expected number of times, out of 98, that a Poisson distribution with a mean of 1.143 would produce a number of 4 or more.

[Or type 'Let c6(5) = 98–31.248–35.717–20.412–7.777' *at the MTB> prompt. Replace c6 and (5) with the column and row you require and the numbers with those appropriate for your data.]*

7 You must amalgamate the observed frequencies in exactly the same way as the expected. In this case there are a total of four observations of 4 or more lice per fish. You should now have one column of expected frequencies with no values less than one and one of observed.

In the lice and fish example the following columns should result

C6 Expected	C7 Observed
31.2484	37
35.7169	32
20.4122	16
7.7771	9
2.8460	4

8 Finally we reach the chi-square test itself. There is no way to reach the required test using the menus. You will have to make sure the command line is enabled (from the 'Editor' menu select 'Enable commands'). Type at the command line (MTB>) the following: 'LET K1 = SUM((C7 – C6)**2/C6)' (assuming that your expected values are in C6 and observed in C7) followed by 'PRINT K1' to see the result. For the example this will return the value of '3.059' (slightly different numbers will be the result of rounding in the calculations).

9 Finally, to determine whether this is significant or not (i.e. do we reject the null hypothesis that the lice are attacking the fish at random) we need to be sure on the degrees of freedom (d.f.). In the example there were five possible values (so 4 d.f.) but the mean value of the Poisson distribution we are comparing against was taken from the data we are using so we lose a further degree of freedom (leaving 3 d.f.). You can either look up the value on a table in a statistics book or you can use the command line in MINITAB using the following commands:

```
MTB > cdf k1 k2;
SUBC> chisquare 3.  (replacing 3 with the degrees of freedom required)

MTB > let k3=1-k2
MTB > print k3
```

This will give you the output:

```
Data Display
  K3   0.382556
```

This value is well above the critical 0.05 level. Therefore we have no reason to reject the null hypothesis that lice attach fish at random in this population. If you look up the value in a statistics book you will see that the chi-square value required to reach the critical level of 0.05 for 3 d.f. is above 7 and here the Chi-square value was only just above 3.

Excel   Most of the calculation steps are fairly straightforward and therefore ideal for a spreadsheet like Excel. Using the same example as for SPSS (above) I will assume that the data are available as a frequency table rather than a column containing all 98 observations. If they are not then you can generate a frequency table from the raw data using the 'Frequency' command (or by hand). The number of lice is in column A and the frequencies are in column B.

1 To calculate expected values for a Poisson distribution the only parameter you need to know is the mean. To find the mean of data in a frequency table you first need to find the product (what you get when you multiply the category value by the number of observations). If you have input the categories in column A with a title, number of observations in B with a title then go to cell C2 and type '=a2*b2'. In this example this will be zero. Then find all the other products by dragging the little square in the bottom right corner of the cell down to the bottom of the list.

Add up the number of observations in the B column by typing 'SUM(b2:b9)' in cell B11 and dragging this across to the C column to add up the products. Then divide the products by the number of observations to obtain the mean by typing '=c11/b11' in cell C13. Always use labels in other cells to make everything clear. In this example the mean should be 1.14.

2 Calculation of the expected number of observations is surprisingly easy in Excel. Move to cell D2 (for convenience) and click on 'Insert function'. Select category 'Statistical' and then 'POISSON' from the very long list of options you are offered; click 'OK'. A window with three lines to fill appears. In the first 'X' line use the cell number of the category, in this case A2. In the second type in the mean (1.143); don't use the cell number. Type a zero in the bottom line. This tells Excel that you do not want the cumulative probabilities. Then click on 'OK'. The probability of getting a zero in a Poisson distribution with a mean of 1.143 appears in the selected cell (just under 0.32).

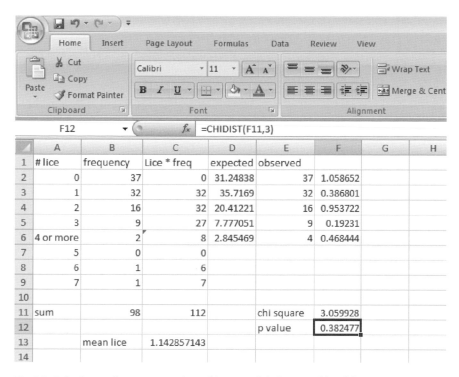

Fig. 7.3 Calculating the mean number of lice per fish from a table of frequencies in Excel. (*Note*: the numbers and letters around the sides of the data are the Excel row and column labels.)

3 Step 2 calculated the probability of getting a particular number of lice. What is needed is the expected *number* of observations. To get this you must multiply the probability by the number of observations, 98 in this example. To do this make sure that cell D2 is selected and add a '*98' to the end of the formula, giving: '=POISSON(A2,1.143,0)*98'. Or, to make the calculation more versatile, use the cell where the total number of observations is calculated (b11 in the example, so b11). Press return and the number in the cell becomes the expected number of observations. Remember to replace 1.143 with your mean and 98 with the number of observations in your sample or use cell numbers anchored with $.

4 Select D2 again and drag its contents down the column (click on the little square in the bottom right corner of the cell and hold the mouse button as you move mouse to highlight all cells to D9 then let go). The column fills with the expected number of observations for each of the categories in column A. Label column D 'expected'. See Fig. 7.3.

5 Some of the expected values are less than one and need to be amalgamated. In this case it is best to have a category for lice numbers of 4 and above. Copy the observed data into column E but replace the number of times four lice were observed with the number of times four or more lice were observed. This is a total of four observations in this case. Label column E 'observed'.

6 You also have to amalgamate the expected frequencies so that cell D6 holds not the expected number of fours but the expected number of '4 or mores'. There are several ways of achieving this but probably the easiest is to start with the 98 total and take off the expected numbers of 0, 1, 2 and 3. Type in cell D6 '=98–D2–D3–D4–D5'. This should give you an expected value of 2.845.

7 Chi-square uses the formula: '(observed–expected)²/expected' for each category and then sums these to produce the final chi-square statistic. The formula is often quoted as $(O–E)^2/E$. In this example the observed values should be in column E and the expected in column D. So in cell F2 type in the formula: '=(E2–D2)^2/D2'. Once you have done that copy it down the F column to F6.

8 You should have five numbers in column F. These need to be added up to give the final chi-square value. Type in any clear cell: '=sum(F2:F6)'. For convenience I used cell F8. You should get the value: 3.0599. This is the chi-square value that you would quote in a report.

9 Finally, what is the probability of getting this value of chi-square (or a higher value)? Go to any clear cell. Click on 'Paste function' and select 'CHIDIST' from the list (if it does not appear immediately it can be selected from either the 'Statistical' list or the 'All' list). In the first, 'X', box you input the chi-square value (or the cell that you used in step 8). In the second box you need to put the degrees of freedom. In this case there were five categories, giving four degrees of freedom, but the Poisson distribution we are comparing against has its mean taken from the frequency data, which loses another degree of freedom (Fig. 7.3). Therefore you should have three degrees of freedom. Input three and you should get a probability of 0.38. This value is well above 0.05 so you can infer that the distribution of lice on fish is not significantly different from random (Poisson distribution). We have no reason to reject the null hypothesis that lice attack fish at random in this population.

Kolmogorov–Smirnov test

The Kolmogorov–Smirnov test for goodness of fit has a variety of uses for large samples of continuous data. There are two main forms, called the one-sample test and two-sample test. Both are used to compare two sets of data to determine whether they come from the same distribution. The one-sample version is more commonly used and compares experimental data with expected distributions. The expected distribution may be derived from the data or may be completely independent of them. For example, you may use the test to determine whether a set of tarsus-length data differs from a normal distribution with the same mean and variance as the sample data before you use parametric analysis on it. The two-sample test can be used to compare a set of egg-weight data from a population of ducks with a set from another site, asking whether the distributions are the same.

> *Note*: although the Kolmogorov–Smirnov test appears similar to the t-test and the Mann–Whitney U test it is not aimed at the same question. The Kolmogorov–Smirnov test delivers a probability that two distributions are the same while the t-test is concerned with means and the Mann–Whitney U test with medians. Two distributions may have identical means and medians and yet have differences elsewhere in their distributions.

An example

The weight in grams is recorded for a sample of 48 mice. This sample is part of an experiment and the researchers wish to know whether the weights are distributed normally before they go on to use parametric statistics. The data are shown here.

12.5	13.5	13.2	12.5	12.1	12.6	12.1	12.8
14.2	13.2	13.8	12.0	12.5	12.1	12.8	12.9
12.6	12.8	12.5	13.1	12.4	13.5	13.4	13.6
13.0	14.1	12.6	13.2	13.8	13.8	13.9	14.0
14.1	12.1	12.9	14.5	13.2	14.1	12.5	12.5
15.0	12.6	13.0	13.5	14.0	12.9	12.4	12.8

SPSS This is a very simple test to access in SPSS. Ensure all the data are in a single column. Select the 'Analyze' menu, then 'Nonparametric Tests', then '1-Sample K–S…'. In the dialogue box move the name of the column you are testing into the 'Test Variable List:' box. Make sure that the 'Test distribution' has 'Normal' selected. Then click 'OK'. The following output will appear:

NPar Tests

One-Sample Kolmogorov-Smirnov Test

		mouse_wt
N		48
Normal Parameters[a,b]	Mean	13.108
	Std. Deviation	.7178
Most Extreme Differences	Absolute	.115
	Positive	.115
	Negative	−.082
Kolmogorov-Smirnov Z		.795
Asymp. Sig. (2-tailed)		.552

a. Test distribution is Normal.
b. Calculated from data.

This output confirms the test and variable used, gives statistics: number of observations ('N'), mean and standard deviation. The last two lines of the output table refer to the test itself. The only important bit is the 'Asymp. Sig. (2-tailed)'. If this number is less than 0.05 then the distribution of the data is significantly different from normal. In this case the value is 0.552 which is well above the critical 0.05 so we have no reason to suppose that the distribution of the data is significantly different from a normal distribution.

R The Kolmogorov–Smirnov test function in R is 'ks.test()', but as the Kolmogorov–Smirnov test has several incarnations available in R it is important to specify carefully exactly what test is to be carried out. In this example the data is held in 'var1' and it is being compared with a normal distribution with the same mean and standard deviation. So R is asked to derive and use those values:

```
> ks.test(var1, "pnorm", mean=mean(var1),
sd=sqrt(var(var1)))

One-sample Kolmogorov-Smirnov test

data: var1
D=0.1147, p-value=0.5523
alternative hypothesis: two-sided

Warning message:
In ks.test(var1, "pnorm", mean=mean(var1),
sd=sqrt(var(var1))) : cannot compute correct p-values
with ties
```

The output gives the output statistic 'D' and then a *P*-value associated with that statistic. Here it is 0.5523 which is well above 0.05, so we accept the null hypothesis that there is no difference between the observations and a set of random observations drawn from a perfect normal distribution with the same mean and variance. There is a warning message that appears often with ranked tests when there are tied values (i.e. two or more identical values in the data set). You should not be concerned unless there are many tied values, in which case the data may not be appropriate for a ranked test, or the *P*-value is very close to 0.05, where you might want to measure with sufficient precision to remove the ties.

MINITAB The Kolmogorov–Smirnov is a very simple statistic to reach if you are testing a single column to see whether it follows a normal distribution or not, although the Anderson–Darling test, considered below, is easier to reach.

Ensure all the data you wish to test are in a single column. Select the 'Stat' menu, then 'Basic statistics...', the 'Normality Test...'. In the dialogue box move the column you wish to test into the 'Variable:' box and select the Kolmogorov–Smirnov test. Then click 'OK'.

The output is a graph that shows a perfect normal distribution of data as a straight line and your data as a series of dots. The test compares the dots with the line. A box next to the graph has the basic statistics of the data, a value for the Kolmogorov–Smirnov test and then a *P*-value. If the *P*-value is less than 0.05 then the distribution is significantly different from normal. In this case the value is given as >0.15 so it is not significantly different from normal.

The Ryan–Joiner test for normality and the Anderson–Darling test are also offered in this dialogue box. Be warned that different normality tests will often give quite different results.

(*Or, if the command interface is enabled, at the MTB> prompt in the session window type* 'NormTest C1;' *then at the SUBC> prompt type* 'KSTest.' *Or you can input commands using* 'Edit' *menu then* 'Command Line Editor'.)

Excel There is no direct method for performing the Kolmogorov–Smirnov in Excel.

Anderson–Darling test

The Anderson–Darling test is one of many procedures commonly encountered to test whether a set of data follows a normal distribution or not. The *P*-value reported is the probability of the data being normally distributed. If $P < 0.05$ the data deviated significantly from a normal distribution and parametric tests should not be used without making suitable corrections or transformations.

SPSS The Anderson–Darling test is not available in this package.

R The Anderson–Darling test is not easily available in R.

MINITAB There are two routes to the Anderson–Darling in MINITAB. The first is described above for the Kolmogorov–Smirnov test. The Anderson–Darling test is also part of the extensive 'Graphical summary' output for simple variable descriptions. The data should be in a single column. From the 'Stat' menu select 'Basic Statistics' then 'Graphical summary'. Move the name of the column containing the data into the 'Variables:' box. Click 'OK'.

An output containing several graphs and many descriptive statistics appears. The first part of the numerical output give the statistics for the Anderson–Darling test ('A-squared') followed by the *P*-value associated with the statistic. If the *P*-value is less than 0.05 the data are significantly different from a normal distribution and it is inadvisable to use parametric statistics.

Excel Tests of normality are not available in this package.

Shapiro–Wilk test

This is another commonly encountered normality test.

SPSS The Shapiro–Wilk test is not available in this package.

R This is the easiest-to-access normality test in R, here assuming that the data are in a variable 'V1':

```
> shapiro.test(V1)

Shapiro-Wilk normality test

data: V1
W=0.9533, p-value=0.05407
```

This indicates that the null hypothesis is very close to rejection and caution should be taken when proceeding with tests that make an assumption of a normal distribution. Note that the Kolmogorov–Smirnov test on the same data gave a *P*-value of 0.5523, well away from borderline significance.

MINITAB The Shapiro–Wilk test is not available in this package.

Excel Tests of normality are not available in this package.

Graphical tests for normality

It's often a good idea to use a visual fit of data to a normal distribution to confirm the results of one of the goodness of fit tests described above.

SPSS An easy graphical way to compare a data set to a distribution is to generate a histogram and then add the normal curve with the same mean and standard deviation as the data. From the 'Graphs' menu select 'Chart Builder...'. In the 'Gallery' tab select the 'Histogram' option and then drag the simple histogram to the 'Chart preview' area. Drag the name of the variable you want from the 'Variables:' list to the 'X-Axis?' section of the preview chart. Now click the 'Element Properties...' button and in the window that appears click 'Display normal curve' and then the 'Apply' button. Close the window and click 'OK' in the 'Chart Builder' window.

Note that if you want a different distribution, SPSS can provide many options. Edit the chart in the 'Output' window by double clicking on it. Remove the existing distribution line, then use the 'Elements' menu to select 'Distribution curve' and you will be offered a choice of lines.

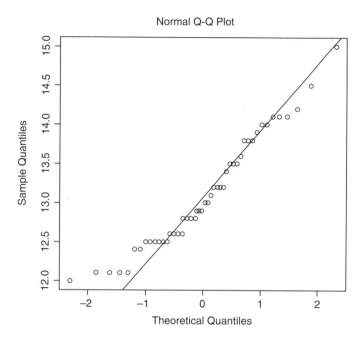

Fig. 7.4 A Q-Q plot produced in R. (Note that where there are several observations with the same value this forms a horizontal row of symbols.)

R An easy way to visualize a fit is to use a 'quantile-quantile plot' with plots the proportion of observations of a set distribution on the horizontal axis against the observed data on the vertical axis (see Fig. 7.4). In R the function 'qqline()' takes the mean and standard deviation from the variable stated. In this case the mouse-weight data is in variable 'V1':

```
> qqnorm(V1)
> qqline(V1)
```

Another simple method in R superimposes a normal curve with mean and variance taken from the data on a histogram of the observations. A cumulative density plot of the observations and a normal distribution can be achieved by first setting a suitable range of values that will need to be plotted and putting them into a variable 'x':

```
> x<-seq(11.5, 15.5, 0.1)
```

where 'seq()' gives the starting point, end point and the spacing. Next make a cumulative density plot of the data in V1 and then use 'lines()' to add a cumulative normal distribution with the mean and standard deviation taken from the observations and plotted at all points of x:

```
> plot(ecdf(V1))
> lines(x, pnorm(x, mean=mean(V1), sd=sqrt(var(V1))),
lty=3)
```

or more tidily as:

```
> plot(ecdf(x), do.points=FALSE, verticals=TRUE)
> lines(x, pnorm(x, mean=mean(V1), sd=sqrt(var(V1))),
lty=3)
```

where '`lty=3`' simply described the pattern of the line.

MINITAB   There are several graphical methods available in MINITAB. For a simple histogram with a normal distribution superimposed on it go to the 'Stat' menu, then 'Basic Statistics...' then 'Graphical Summary' and the histogram of the data is the first part of the output. To get a window with only the histogram and normal curve go to 'Stat' menu, then 'Basic Statistics', then 'Display Descriptive Statistics...'. Move the variable of interest into the 'Variables:' box then click on 'Graphs' and select 'Histogram of data, with normal curve'.

Excel   No simple graphical methods for testing normality are available in Excel.

Do the observations from two groups differ?

The two groups can be paired, repeated or related samples or they can be independent. Paired measures are considered first.

Paired data

Paired samples or paired comparisons (paired data; a.k.a. related or matched data) occur when a single individual is tested twice (e.g. before and after) or a sampling station retested. Another possible use occurs when an individual is, or individuals of a clone are, divided and then subjected to two treatments. Three tests are considered below: the **paired *t*-test, Wilcoxon signed ranks test** and the **sign test**.

Paired t-test

The data must be continuous and, at least approximately, normally distributed. The variances of the two sets must be homogeneous (this can be tested by the Levene test). The null hypothesis is that the there is no difference between the two columns and they could come from the same data set.

An example It is suggested that the building of a power station will affect the amount of particulate matter in the air. However, there are only three readings available for the month before the project got underway (measured as parts per million, or ppm). The sites where the three readings were taken were revisited once the station was complete.

Site	Before	After
1	34.6	41.3
2	38.2	39.6
3	37.6	41.0

SPSS Arrange the data into two columns of equal length such that each row represents one individual (or site in this case). The columns should be labelled 'before' and 'after' as this will make it easier to interpret the output.

Under the 'Analyze' menu choose 'Compare Means' and then 'Paired-Samples T Test...'. In the dialogue box that appears move 'Before' into 'Variable1' and 'After' into 'Variable2' in 'Pair' row 1. Then click 'OK'.

The first part of the output gives some information about the data like this:

Paired Samples Statistics

		Mean	N	Std. Deviation	Std. Error Mean
Pair 1	Before	36.800	3	1.9287	1.1136
	After	40.633	3	.9074	.5239

Paired Samples Correlations

		N	Correlation	Sig.
Pair 1	Before & After	3	−.749	.462

It tells you the names of the two variables, means, how many observations there were, standard deviation and standard error of the two sets of data. The next table refers to a Pearson product moment correlation done on the data, with the value of r given as 'Correlation' and 'Sig.' being the test of the null hypothesis that r is 0 (i.e. no correlation).

The second part of the output is the report from the t-test itself:

Paired Samples Test

	Paired Differences							
			Std. Error Mean	95% Confidence Interval of the Difference		t	df	Sig. (2-tailed)
	Mean	Std. Deviation		Lower	Upper			
Pair 1 Before - After	−3.8333	2.6764	1.5452	−10.4820	2.8153	−2.481	2	.131

The first three columns refer to the mean difference between pairs of data, followed by the standard deviation and standard error of the differences. Next comes the confidence interval of the mean saying it is 95% likely to be between the two values given. The 95% CI is very wide because there are only three pairs of data. Finally comes the *t*-test itself with the result of the test (the value given can be looked up on a Student's *t*-table). Then the degrees of freedom (number of pairs minus 1) and the probability of this *t* value (or larger) occurring if the null hypothesis is correct. In this case the probability is 0.131 (or 13.1%) and therefore we conclude that there is no significant difference between the two columns. However, with such a small set of data, achieving a significant result is extremely unlikely.

R In most cases you will read in the data from a file and attach it. Here, as there are very small samples, I'm typing the data in directly into variables (vectors) called 'before' and 'after', then carrying out a paired *t*-test, which uses the R function 't.test()' with the option 'paired=T' to indicate that it's a paired test.

```
> before=c(34.6,38.2,37.6)
> after=c(41.3,39.6,41.0)
> t.test(before,after,paired=T)

Paired t-test

data: before and after
t=-2.4807, df=2, p-value=0.1313
alternative hypothesis: true difference in means is not
equal to 0
95 percent confidence interval:
-10.481980 2.815313
sample estimates:
mean of the differences
-3.833333
```

The output confirms that a paired *t*-test is carried out and gives the names of the two variables. The *t* statistic, degrees of freedom and *P*-value are given. The output then reminds you that what is being tested is whether the difference between the two variables is significantly different from zero. It then gives the upper and lower values of the 95% CI of the difference. If the 95% confidence values both have the same sign, then the mean difference between the two samples is significantly different from zero. Here, with such small sample sizes, even with a mean difference of −3.83 the confidence intervals straddle zero.

MINITAB First input the data into two columns, one for before and the other after. There is no need to have a separate column to label the individual sites although this might help you interpret the results.

1 From the 'Stat' menu, select 'Basic Statistics' and then 'Paired t...'. Move the two variables into the boxes labelled 'First sample:' and 'Second sample:'. Click 'OK'.

You get the following output from the example.

Paired T-Test and CI: Before, After

```
Paired T for Before - After

                N      Mean     StDev    SE Mean
Before          3      36.80     1.93      1.11
After           3      40.63     0.91      0.52
Difference      3      -3.83     2.68      1.55

95% CI for mean difference: (-10.48, 2.82)
T-Test of mean difference = 0 (vs not = 0): T-Value = -2.48 P-Value = 0.131
```

---

2 The output confirms that you have run a paired t-test of before and after. There are three pairs of data in the example giving a value of 3 for N with summary statistics. A line on the table of 'Difference' is important as the test is actually comparing the values of difference against zero. Finally is shown a t value and then the important value, the P-value. If this is less than 0.05 then you must reject the null hypothesis that the difference between before and after is zero. In this case the value of 0.131 indicates that there is not enough evidence in the three pairs of data to reject the null hypothesis. However, with such a small data set in the example it is not surprising that you can't detect a significant effect.

3 The null hypothesis is that the differences are not significantly different from zero. If you want to test a difference other than zero, or you want to have a one-tailed test (i.e. you are only interested in the hypothesis that 'before' is greater than 'after', with the null hypothesis that 'before' is *not* greater than 'after') then go to 'Options' in the 'Paired-t' dialogue before running the test.

Excel The data should be input in two columns of equal length. If the spreadsheet were set up using the example exactly as the table above, then the data for 'before' would be in cells b2, b3 and b4, while that for 'after' would be in cells c2, c3 and c4. In Excel it doesn't really matter where the data are on the spreadsheet as long as you know the cell numbers.

Select an empty cell where you want the result reported. Then there are two ways of achieving the same result.

1 Go to 'Insert function', select 'Statistical' and then 'TTEST'. Define Array 1 as 'b2:b4' and Array 2 as 'c2:c4'. Select 'Tails' as '2' (you will nearly always require a two-tailed test) and the 'type' as '1' (this selects a paired test in Excel). The probability or *P*-value of '0.131' will then appear in the cell.

2 Type in '=TTEST' followed by the first and last cell of the first column separated by a colon, then the same for the second column. Then the number of tails in the test (usually 2) and then a 1 to ask for a paired test. In this case you would type '=TTEST(b2:b4,c2:c4,2,1)' and the probability will appear in the cell.

This will give you the *P*-value associated with the paired *t*-test. To get the value of the *t* statistic you can use the function 'TINV'. Select a blank cell and type in '=TINV(F8,2)' replacing 'F8' with the cell containing the result in the previous step and replacing '2' with the degrees of freedom which is the number of pairs of observations minus 1.

Wilcoxon signed ranks test

This test is the non-parametric equivalent of the paired *t*-test. It has far fewer assumptions about the shape of the data although it does assume that the data are on a continuous scale of measurement. This means that any type of length, weight, etc. will be suitable. The test is somewhat less powerful than the paired *t*-test. A minimum of six pairs of data are required before the test can be carried out.

An example In this example the 'individuals' are sampling stations in a river system and the data are measures of flow (in litres per second). The investigator wishes to know if the flow is significantly different on the two days. The null hypothesis is that there is no difference in flow.

Station	Day 1	Day 2
1	268	236
2	260	241
3	243	239
4	290	285
5	294	282
6	270	273
7	268	258

SPSS Arrange the data into two columns of equal length such that each row represents one individual. The columns should be labelled.

From the 'Analyze' menu choose 'Nonparametric Tests' and then '2 Related Samples...'. As a default the 'Wilcoxon' test should be checked but ensure this is the case. Move the two variables ('Day_1' and 'Day_2' in this example) to the first row in the 'Test Pairs:' box with the right arrow then click 'OK' to run the test.

The output will look like this:

Wilcoxon Signed Ranks Test

Ranks

		N	Mean Rank	Sum of Ranks
Day_2 - Day_1	Negative Ranks	6[a]	4.50	27.00
	Positive Ranks	1[b]	1.00	1.00
	Ties	0[c]		
	Total	7		

a. Day_2 < Day_1
b. Day_2 > Day_1
c. Day_2 = Day_1

Test Statistics[b]

	Day_2 - Day_1
Z	−2.197[a]
Asymp. Sig. (2-tailed)	.028

a. Based on positive ranks.
b. Wilcoxon Signed Ranks Test.

The test classifies the paired data into three categories: those where 'Day_2' is less than 'Day_1' ('Negative Ranks'); those where it is greater ('Positive Ranks') and those where they are the same ('Ties'). In this case there are six of the first, one of the second and no ties. It also ranks the absolute differences (from smallest difference as rank one to the largest as rank seven). In this case the one pair where the flow is greater on 'Day_2' is also the smallest difference, as it is ranked '1.00'. The output value of 'Z' in the second table may be looked up in statistical tables but the result is given anyway as a 'Asymp. Sig. (2-tailed)'. In this case $P<0.05$ so the null hypothesis must be rejected. The alternative hypothesis that the flows were different on the two days is accepted.

R Assuming the data have been input or imported and the two vectors containing the data are in 'Day1' and 'Day2' you can then use 'wilcox.test()'. Here we specify 'paired=TRUE' and this means the test will be comparing the differences between the two observations against a null hypothesis that the differences have a median of zero.

```
> wilcox.test(Day1, Day2, paired=TRUE)

Wilcoxon signed rank test

data: Day1 and Day2
V=27, p-value=0.03125
alternative hypothesis: true location shift is not
equal to 0
```

Here the *P*-value is less than 0.05, so we reject the null hypothesis that the median difference between 'Day1' and 'Day2' is zero.

Note that R will compute what is called an 'exact' version of the test to calculate the *P*-value if there are fewer than 50 pairs of observations. This will have a small impact on the *P*-value in most cases, although to retain consistency between tests where some are above 50 and some below, you can use 'exact=TRUE' or 'exact=FALSE' as an option within the 'wilcox.test()'. In this example 'exact=F' will alter the *P*-value to 0.03461, which is the value given in MINITAB.

MINITAB This test is not achievable in a single step. However, if the data are arranged in two columns you can carry out an analogous test to the paired *t*-test.
1 You need to create a new column that contains the difference between the first observation and the second. Go to the 'Calc', select 'Calculator...'. Type 'Diff' in the 'Store result in variable:' and 'Day 1' – 'Day 2' (either by typing or double-clicking) in the 'Expression:' box (replacing with the names of your variables as appropriate).
2 The null hypothesis is that the median of the differences is not significantly different from zero. This is tested using a one-sample Wilcoxon test. Go to the 'Stat' menu, then 'Nonparametrics', then '1-sample Wilcoxon....'. In the dialogue box put 'Diff' in the 'Variables:' box. Select the 'Test median' option, leave the value as 0.0 and make sure that the 'not equal' option is selected from the pull-down menu. Click 'OK'.

You get the following output from the example.

Wilcoxon Signed Rank Test: Diff

```
Test of median = 0.000000 versus median not = 0.000000

                N for      Wilcoxon                  Estimated
         N      Test       Statistic        P         Median
Diff     7       7            27.0        0.035        10.50
```

---

3 The output confirms that you are testing the null hypothesis that the median of 'Diff' (i.e. the difference between the flows on the two days) is zero (with the package deciding to give a ridiculous number of decimal places). The important value is the *P*-value 'P', which is 0.035 in this example. This is less than the critical 0.05 and the null hypothesis should be rejected. The alternative hypothesis H_1 is that the median of the differences is not equal to zero. This means that

the flows on day 1 and day 2 were different. Inspection of the raw data or the estimated median shows that flow was greater on day 1.

Excel There is no easy method for performing the Wilcoxon signed ranks test in Excel.

Sign test

This is a very simple test that makes almost no assumptions about the form of the data, only that it is possible to compare them in some way to decide which is larger. The test is of very low power but very safe (i.e. it is a conservative test and type I errors are very unlikely). The sign test should only be used when there are large numbers of paired observations.

The test works using the assumption that if two sets of observations are not different then there will be the same number of pairs when A is bigger than B as there are when B is bigger than A. Therefore the actual values of the data points are relatively unimportant as long as they can be compared to see which is the larger. This means the test is not very sensitive to poor quality of data.

The sign test uses the binomial distribution (page 37) to calculate significance values.

An example The same data as was used as for the Wilcoxon signed ranks test will be used in this example.

SPSS Arrange the data into two columns of equal length such that each row represents one individual (or site). The columns should be labelled appropriately.

From the 'Analyze' menu choose 'Nonparametric Tests' and then '2 Related samples'. As a default the 'Wilcoxon' test should be checked, uncheck this and check 'Sign' instead. Select the two variables ('Day_1' and 'Day_2' in this example), move them to the first row of the 'Test Pairs:' and click 'OK'.

The output using the example data (described above for the Wilcoxon signed ranks test) will look like this:

Sign Test

Frequencies

		N
Day_2 - Day_1	Negative Differences[a]	6
	Positive Differences[b]	1
	Ties[c]	0
	Total	7

a. Day_2 < Day_1
b. Day_2 > Day_1
c. Day_2 = Day_1

Test Statistics[b]

	Day_2 - Day_1
Exact Sig. (2-tailed)	.125[a]

a. Binomial distribution used.
b. Sign Test

For the seven pairs of observations there are six where 'Day_2' is less than 'Day_1' ('Negative differences'). The probability of this or a more extreme result occurring by chance is 0.125 or 12.5% and this is given as the 'Exact Sig. (2-tailed)'. In fact it is giving the probability of getting zero, one, six or seven out of seven the same direction, given that we are expecting the chance of getting a higher or lower value to be 0.5 (i.e. equal chance of A bigger than B as for B bigger than A).

In this case we would not reject the null hypothesis that the two days had different flows. This shows clearly that the sign test is of very much lower power than the Wilcoxon signed ranks test. In fact, the sign test is really only useful as the number of paired observations becomes quite large.

R There is no direct way to carry out this test in R without downloading a package. However, it is a very simple test that makes use of the binomial distribution. First make sure the data are available in two labelled vectors. Here I've used the example from the Wilcoxon signed ranks test and have two sets of seven observations in vectors labelled 'Day1' and 'Day2'.

We then need to know how many pairs of observations there are ('length()') and how many times the value in 'Day1' is greater than that in 'Day2':

```
> length(Day1)
[1] 7
> sum(Day1>Day2)
[1] 6
```

The probability of getting a value this extreme or more extreme is calculated in a binomial test:

```
> binom.test(6,7)

Exact binomial test

data: 6 and 7
number of successes=6, number of trials=7,
p-value=0.125
alternative hypothesis: true probability of success is
not equal to 0.5
```

```
95 percent confidence interval:
0.4212768 0.9963897
sample estimates:
probability of success
0.8571429
```

Here the *P*-value is greater than 0.05, so we can't exclude the null hypothesis that there is an equal probability of 'Day1' being greater than 'Day2' and 'Day1' being less than 'Day2'. The 95% confidence interval of the prediction crosses the null value of 0.5 (i.e. equal probability of the two outcomes). In fact this test is very weak and is unlikely to get a significant result with such a small number of observations. For a sample size of seven only seven out of seven or zero out of seven will give a *P*-value below 0.05.

In R you can easily combine all the steps above into a single line:

```
> binom.test(sum(Day1>Day2),length(Day1))
```

You must make sure that there are the same number of observations in each vector and that they are lined up properly (i.e. each element in a vector is paired with the corresponding element in the other vector).

MINITAB This test is carried out in a very similar way to the Wilcoxon signed ranks test. Arrange the data in two columns. Label the columns appropriately.
1 Create a new column that contains the difference between the first observation and the second. Go to the 'Calc', select 'Calculator...'. Type 'Diff' in the 'Store result in variable:' and 'Day 1' – 'Day 2' (either by typing or double-clicking) in the 'Expression:' box (replacing with the names of your variables as appropriate). Obviously, you will have to replace these names with the appropriate names for the data columns.
2 The null hypothesis is that there are equal numbers of positive and negative differences. This is tested using a one-sample sign test. Go to the 'Stat' menu, then 'Nonparametrics', then '1-sample Sign...'. In the dialogue box put 'diff' in the 'Variables:' box. Select the 'Test median' option, leave the value as 0.0 and make sure that the 'not equal' option is selected from the pull-down menu. Click 'OK'.

You get the following output from the example.

Sign Test for Median: Diff

Sign test of median = 0.00000 versus not = 0.00000

	N	Below	Equal	Above	P	Median
Diff	7	1	0	6	0.1250	10.00

---

3 The output confirms that you are testing the null hypothesis that the median of 'diff' (the difference between the flows on the two days) is zero. The important value is the *P*-value 'P', which is 0.1250 in this example (i.e. there is a 12.5% chance of getting this result or a more extreme one). This is greater than the critical level of $P=0.05$ and the null hypothesis should be accepted. This is the same data as was used in the Wilcoxon signed ranks test that rejected the null hypothesis, thus demonstrating what a conservative test it is.

Excel There is no direct method for the sign test in Excel. However, if you compare each pair of data and count up the number of times the first observation is greater then you can use the 'Binomdist' function in Excel to work out the *P*-value. The binomial distribution is described in Chapter 5. Here I give two methods for the sign test in Excel and a method for counting up the number of '+' and '−' pairs.

Using the example above, there are seven pairs of data. In six of the seven pairs the first column is greater. We are assuming a null hypothesis that there is no difference between the two columns. If that is the case then we expect an equal number of observations where the first column is greater and when the first is smaller.

1 In Excel the probability of getting six plusses out of seven can be determined fairly easily. After selecting an empty cell where you want the result to be displayed there are two methods.

Either: use 'Insert Function', select 'Statistical', then 'BINOMDIST', click 'OK' and you will be confronted with four boxes. In the first type the number of successes ('Number_s'), in this case 6. In the second the number of 'Trials', in this case 7. Next the probability of getting a 'success', in this case we choose 0.5 (in other words, we assume an equal chance of getting a plus or a minus). In the last box type 0 to indicate that you don't require the cumulative probability. Click 'OK' and this will give you the result 0.054.

Or: type in the values described above directly: '=BINOMDIST(6,7,0.5,0)'.

This is not the complete answer. It only tells you the chance of getting six out of seven. What you need from the sign test is the chance of getting a result as extreme as six out of seven or more extreme. For seven trials that means six or seven as well as zero or one because the test is two-tailed.

2 In a new empty cell for each possibility, repeat the previous commands but replacing the six successes with seven, zero and one. A good way to do this is to put the four required numbers into a column then use the cell number instead of typing in the number. For example, if cell A1 has a 7 in it, A2 has 6, etc. then just type into cell B1 '=BINOMDIST(A1,7,0.5,0)' and then copy this down the next four cells by clicking on the small black square in the bottom right corner of the cell and dragging it down to cover the cells required.

3 Total up the four probabilities by typing '=SUM(B1:B4)' in an empty cell. This should give you the answer 0.125. This can be interpreted as a 12.5% chance of getting a result as extreme as six out of seven or more extreme. We have to accept the null hypothesis that there is nothing happening between the two sampling events because the value is well above $P=0.05$. This demonstrates

the low power of the test. In fact, with a sample size this small you need to get seven out of seven or zero out of seven to get a significant result (that has a probability of 0.016).

An alternative method in Excel, useful with larger samples, is to use the cumulative probability option. As this is a two-tailed test with a 0.5 chance of plus or minus, the probability of getting one out of seven is the same as getting six out of seven. So to test the probability of getting six out of seven or more extreme is equivalent to testing that of one out of seven. The last number of the 'BINOMDIST' function can be changed from zero to one to give cumulative probability (i.e. the chance of getting that number or lower). So the probability of getting zero *or* one is given by '=BINOMDIST(1,7,0.5,1)'. Doubling this value will give both the probability of getting either six events out of seven or one out of seven.

These methods require that the number of '+' and '−' observations are known. In Excel it is easy to compare two values and tally the number of times A is larger than B using the 'IF' function. Return to the data table with paired values in columns B and C. Select cell D2. Click on 'Insert function', then 'Logical' and 'IF'. The 'Logical_test' is B2>C2, testing whether the value in column B is greater than that in column C. The 'Value_if_true' should be 1 and 'Value_if_false' 0. Click 'OK' and paste this down the D column. A column of six 1s and one 0 should appear. The number of 1 values can be quickly totalled using 'SUM(D2:D8)'. A word of caution. Ties cause problems with this method. I suggest that they are totally ignored in the sign test. So add up the number of times A is bigger than B and then remove all instances when A = B before calculating the total number of observations.

Unpaired data

Unpaired samples or unpaired comparisons occur when a single individual is measured or tested only once. There will, therefore, be two totally separate groups of observations making up the two samples. Two groups are often obvious; for example males and females, or kudu and eland. However, the distinction between the groups may be rather arbitrary, such as eastern and western, or large and small. Three tests are considered below: the **independent-samples *t*-test**, a **one-way analysis of variance** and the **Mann–Whitney U-test**.

t-*test*

The independent-samples *t*-test is the more usual form of the *t*-test and if the term '*t*-test' is not qualified then this is what is being referred to. The null hypothesis is that the two sets of data are the same. (Actually the null hypothesis is that the two sample means come from a population with the same true mean, μ.) The *t*-test assumes that the data are continuous, at least approximately normally distributed and that the variances of the two sets are homogeneous (i.e. the same). If possible these assumptions should be tested before the test is carried out, although this test is often incorporated into the test in statistical packages. If the two sets of observations do not have the same variance then there are ways to adjust the result of the *t*-test to compensate. It is these adjustments are often incorporated into the test in the statistical package.

An example The weights of five grains have been measured from each of two experimental cultivars called *Premier* and *Super*. Each grain has been weighed to the nearest 0.1 mg. The researcher wishes to determine whether the grain weight is the same in the two cultivars. The null hypothesis (H_0) is that the two cultivars have the same mean grain size. The alternative hypothesis (H_1) is that the two cultivars have different mean grain size.

Premier	*Super*
24.5	26.4
23.4	27.0
22.1	25.2
25.3	25.8
23.4	27.1

SPSS Input all of the observed data into a single column. Use another column for the labelling of the groups. This may seem wasteful but it is a much easier system when it comes to multiway analysis where each item of data will belong to several groups simultaneously.

The five grains from each of two cultivars of crop plant have been placed in the second column. The two cultivars have been coded as '1' or '2' in the first column. Columns have been labelled in the 'Variable View'. The package only allows restricted labels so the ideal label 'Grain size (mg)' has been shortened to 'Grain_sz', although the full label can be added using the 'label' column in the 'Variable View'. The names of the groups (cultivars here) can be added using the 'Values' column of the 'Variable View', and the value labels can be shown on the spreadsheet by choosing the 'Value Labels' option under the 'View' menu (see Fig. 7.5).

Under the 'Analyze' menu choose 'Compare Means' and then 'Independent-Samples T Test...'. In the dialogue box that appears move 'Grain_sz' into the 'Test Variable(s):' box by first highlighting it and then clicking on the appropriate move button (a blue right arrow). Next move 'Cultivar' into the 'Grouping Variable:' box. It will appear as 'Cultivar(? ?)'. You need to click the 'Define Groups...' button and then input '1' in 'Group 1:' and '2' in 'Group 2:' before clicking 'Continue'. This will return you to the first dialogue box. You will see that 'Cultivar(? ?)' is now 'Cultivar(1 2)'. Click 'OK' to run the test.

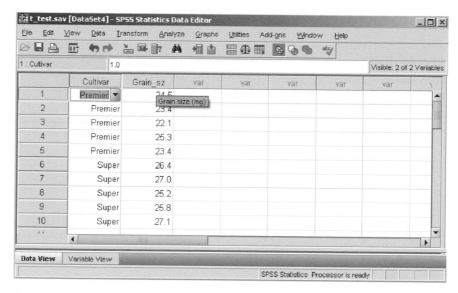

Fig. 7.5 Arranging the data for a *t*-test in SPSS. The groups for 'Cultivar' have been entered as '1' and '2' and then given labels in the 'Variable view'. The selected cell is still filled with the number '1'. Also note that the pop-up label for column two gives the full name of the variable, rather than 'Grain_sz'.

The output will come in two parts. The first gives some information about the data like this:

→ **T-Test**

Group Statistics

	Cultivar	N	Mean	Std. Deviation	Std. Error Mean
Grain size (mg)	Premier	5	23.740	1.2178	.5446
	Super	5	26.300	.8062	.3606

This confirms the type of test. Then in the table shows that the variable used was 'Grain size (mg)' and that there were two groups of 'Cultivar' called 'Premier' and 'Super' each with 'N' of 5 (number of observations). Then come some simple descriptive statistics of the two samples: their mean, 'Std. Deviation' and 'Std. Error Mean' (standard error of the mean). (Note that the full name of the data variable and the group names for '1' and '2' will only appear if they have been added in the 'Variable View'.)

The second part of the output contains the result of the *t*-test itself:

Independent Samples Test

		Levene's Test for Equality of Variances		t-test for Equality of Means					95% Confidence Interval of the Difference	
		F	Sig.	t	df	Sig. (2-tailed)	Mean Difference	Std. Error Difference	Lower	Upper
Grain size (mg)	Equal variances assumed	.762	.408	−3.919	8	.004	−2.5600	.6531	−4.0662	−1.0538
	Equal variances not assumed			−3.919	6.941	.006	−2.5600	.6531	−4.1071	−1.0129

First comes the Levene's test for equality of variances. As the *t*-test assumes that the two samples have equal variance SPSS sensibly tests this every time you carry out a *t*-test. The important figure is the second one. In this case it is 'Sig.=.408' (i.e. $P>0.05$) so there is no evidence that the variances are unequal. If the *P*-value ('Sig.') is lower than 0.05 then you should be wary about using the *t*-test and should consider using the Mann–Whitney U test instead.

The remainder of the output appears on two lines. The upper one, 'Equal variances assumed', is the standard *t*-test and the lower one 'Equal variances not assumed' is a more conservative version that compensates for the possible problems caused by difference in variances by using a reduced value for the degrees of freedom in the test. The lower the 'Sig.' value in the Levene test, the bigger the difference between the two lines. In most cases you should not need to worry about the fact that there are two versions of the test because in most cases, as in this example, they will tell the same story.

The 't' is the actual result of the test that can be looked up on Student's *t*-table. 'df' is the degrees of freedom of the test. This will be two fewer than the total number of observations in the two samples. The result of the Levene's test determines the reduced degrees of freedom used in the lower line.

Next comes the important bit; labelled 'Sig. (2-tailed)', it is the probability that the null hypothesis is correct. This is the *P*-value. In this case the *P*-value is much less than 0.05 so it is clear that the null hypothesis is extremely unlikely to be true. In fact there is only a 0.4% chance that we would see a *t* value as large as this if the null hypothesis is true using the basic *t*-test. This value rises to 0.6% using the more conservative version (on the lower line).

'Mean Difference' gives the value of mean of group 1 minus mean of group 2. It will be a negative value if the mean of group 2 is larger than that of group 1. 'Std. Error Difference' gives the standard error for this difference. Finally comes a '95% Confidence Interval of the Difference' column. This gives the range of difference between the two means within which 95% of samples are likely to come. In the example the 'Lower' and 'Upper' values are both negative which shows that cultivar 2, *Super*, is very likely to be heavier than cultivar 1, *Premier*. When the *t*-test result is not significant the upper 95% confidence value will be positive and the lower negative.

R There are two routes depending on how the data are arranged. In the first the two groups should be in two variables; in the second all the data are in one variable and another variable is used to identify the groups. Both use the function 't.test()' in R.

Method 1 Assuming that the data have been arranged and labelled as in the example use the following command, inserting the names of your variables instead of mine. Although 'paired = FALSE' is the default, so could be removed, it does confirm that this isn't a paired test:

```
> t.test(Premier,Super,paired=FALSE,var.equal=TRUE)

Two Sample t-test

data: Premier and Super
t=-3.9195, df=8, p-value=0.004422
alternative hypothesis: true difference in means is not
equal to 0
95 percent confidence interval:
-4.066158  -1.053842
sample estimates:
mean of x mean of y
23.74       26.30
```

The output confirms the value of *t* and the degrees of freedom 'df'. The *P*-value is well below 0.05 so we reject the null hypothesis that the two sets of data come from distributions with the same mean. The 95% confidence limits both have the same sign here; if they cross zero then the *P*-value will be above 0.05. Finally the mean values of the two variables are given.

The default version of the *t*-test in R doesn't assume that variances are equal and uses the comparison of variances of the two variables to adjust the degrees of freedom. This version of the *t*-test is here called the 'Welch t-test'. This is more conservative, and therefore more likely to generate a type II error and less likely to give a type I error. In the example there is no effect on the conclusions as the *P*-value is still well below 0.05:

```
> t.test(Premier,Super,paired=FALSE)
Welch Two Sample t-test

data: Premier and Super
t=-3.9195, df=6.941, p-value=0.005849
alternative hypothesis: true difference in means is not
equal to 0
```

Method 2 Here the data are in a single variable (vector) and there is another variable that contains the group codes. In the example the data are in 'Grain' and the grouping variable is 'Cultivar'. The syntax in R is used in many functions and states that the model being used is that 'Grain' is a function of 'Cultivar':

```
> t.test(Grain~Cultivar, var.equal=TRUE)

Two Sample t-test

data: Grain by Cultivar
t=-3.9195, df=8, p-value=0.004422
alternative hypothesis: true difference in means is not
equal to 0
95 percent confidence interval:
-4.066158 -1.053842
sample estimates:
mean in group 1 mean in group 2
23.74        26.30
```

Here the output is identical to method 1, except that the data are confirmed to be arranged in a different way. The output confirms the value of *t* and the degrees of freedom 'df'. The *P*-value is well below 0.05 so we reject the null hypothesis that the two sets of observations come from distributions with the same mean. The 95% confidence limits both have the same sign here; if they cross zero then the *P*-value will be above 0.05. Finally the mean values of the two sets of observations are given.

As with method 1, the default method does not assume that variances are equal and uses the comparison of variances of the two variables to adjust the degrees of freedom. This version of the test is more conservative, and therefore more likely to generate a type II error and less likely to give a type I error. In the example there is no effect on the conclusions:

```
> t.test(Grain~Cultivar)

Welch Two Sample t-test

data: Grain by Cultivar
t=-3.9195, df=6.941, p-value=0.005849
```

MINITAB There are two methods. In the first the two groups should be in two columns, and in the second all the data are in one column and another column is used to identify the groups. The second method appears wasteful but it is a required strategy in more complex analysis.

Method 1 Input the data into two columns and label the columns appropriately. From the 'Stat' menu select 'Basic statistics' and then '2-sample t...'. In the dialogue box that appears select the 'Samples in different columns' option. Then place the appropriate column labels ('*Premier*' and '*Super*' in the example) in the two boxes labelled 'First:' and 'Second:'. Make sure the 'Options…' dialogue reads 'Alternative: not equal' and the 'Confidence level:' is set at 95.0 (this means that the mean for the difference between the two means will be calculated with a 95% confidence interval). Leave the 'Assume equal variances' box unchecked. Click 'OK' to run the test.

(Or, if the command interface is enabled, type 'twosample c1 c2' *at the MTB> prompt in the session window. Or you can input commands using* 'Edit' *menu then* 'Command Line Editor'.)

The following output appears:

Two-Sample T-Test and CI: Premier, Super

```
Two-sample T for Premier vs Super


           N    Mean     StDev    SE Mean
Premier    5    23.74    1.22       0.54
Super      5    26.300   0.806      0.36


Difference = mu (Premier) - mu (Super)
Estimate for difference: -2.560
95% CI for difference: (-4.158, -0.962)
T-Test of difference = 0 (vs not =): T-Value = -3.92 P-Value = 0.008 DF = 6
```

This output confirms the test was a *t*-test and confirms the names of the two variables. It then gives summary statistics for the two groups: number of observations ('N'), mean, standard deviation and standard error of the mean. The next line gives the 95% confidence interval for the mean difference between the two groups. The references to 'mu' are to the Greek letter, μ, which is used to denote a mean. In the last line is the output for the *t*-test itself, confirming that a test of equal means is being made. The value of *t* is given as −3.92 and the important *P*-value as 0.008. This value is much less than 0.05 so we reject the null hypothesis and accept the H_1 that the two groups have different means. Inspection of the summary statistics shows that *Super* has larger grain size.

The degrees of freedom ('DF') is given as 6. There should be eight degrees of freedom for the example but, as the assumption of equality of variance was not made, a correction is applied to the degrees of freedom to make the test more conservative. If the 'Assume equal variances' is checked then the example data will give eight degrees of freedom, the same value for *t*, and a *P*-value of 0.004.

Method 2 Input all the data into a single column and use a second column to label the data. These labels should be integers. In the example is probably best to label '*Premier*' as '1' and '*Super*' as '2'. Label the columns appropriately. In the example the data column should be labelled 'Grain sz' and the group codes as 'Cultivar'.

From the 'Stat' menu select 'Basic statistics' and then '2-sample t...'. In the dialogue box that appears select the 'Samples in one column' option. Then place the column with the data in the 'Samples:' box and the one with the group codes in the 'Subscripts:' box. Make sure the pull down menu in the 'Options...' dialogue reads 'not equal' and the 'Confidence level:' is set at 0.95. Leave 'Assume equal variances' unchecked. Click 'OK' to run the test.

(Or type 'twot c1 c2' (assuming data are in c1) at the MTB> prompt in the session window. Or you can input commands using 'Edit' menu then 'Command Line Editor'.)

You get the following output:

```
Two-Sample T-Test and CI: Grain sz, Group

Two-sample T for Grain sz

Group   N    Mean     StDev    SE Mean
1       5    23.74    1.22     0.54
2       5    26.300   0.806    0.36


Difference = mu (1) - mu (2)
Estimate for difference: -2.560
95% CI for difference: (-4.158, -0.962)
T-Test of difference = 0 (vs not =): T-Value = -3.92 P-Value = 0.008 DF = 6
```

Apart from the code names '1' and '2' replacing the group names the output is identical to that produced in Method 1.

Excel In this case the data may be anywhere on the spreadsheet. As long as you know the cell locations of the two groups there is no problem. However, in practice, it is much easier if the data are either input exactly as in the SPSS example above, with one column defining the group and another containing the actual data, or in two adjacent, and clearly labelled, columns.

Assuming you have input the data in an identical format to the SPSS example for the *t*-test then the data for the first cultivar is in cells b2–b6 and for the second cultivar in cells b7–b11. There are now two methods that may be used.

Method 1 Go to 'Insert function', select 'Statistical' and then 'TTEST'. Define the first array ('Array 1') as 'b2:b6' and the second array as 'b7:b11'. These arrays may be defined by selecting the box then clicking and dragging over the appropriate cells on the spreadsheet. Select 'tails' as '2' (you will nearly always require a two-tailed test) and the 'type' as '2' (this selects a standard *t*-test; '1' gives a paired test). Click 'OK'. The probability or *P*-value of 0.0044 will then appear in the cell.

Method 2 Type in 'TTEST' followed by, in parentheses, the first and last cell of the first column separated by a colon, then the same for the second column. Then the number of tails in the test (usually 2) and then a 2 to ask for a standard *t*-test. In this case you would type '=TTEST(B2:B6,B7:B11,2,2)' and the probability will appear in the cell. Excel does not report the *t* value.

If you need the value of *t* for a report, then you can use the function 'TINV' to convert the *P*-value and degrees of freedom back into a *t* value. Using the example, inserting '=TINV(0.0044,8)' into a blank cell will give the result 3.92. Replace the values with those appropriate for your data, or point at the relevant cell on the spreadsheet.

If you are concerned that the variances of the two samples may not be equal, or you know it to be the case, then you should not use the standard *t*-test. Excel allows you to carry out a *t*-test that does not assume homogeneity of variances. This is easily accessed by using a type 3 test instead of a type 2. If you use a type 3 *t*-test with this example the probability should be reported as 0.0058.

One-way ANOVA

Using analysis of variance (ANOVA) to determine whether just two groups have the same mean may seem like overkill. This may be the simplest use of ANOVA but it still works and gives the same answer as the *t*-test. I am of the opinion that the fact that the *t*-test is restricted to two groups makes the use of ANOVA preferable in this situation because you don't have to learn a new test when you consider more than two groups.

ANOVA has the same basic assumptions as the *t*-test: that the data are continuous, at least approximately normally distributed and the variances of the data sets are homogeneous. These assumptions should be tested before the test is carried out. The null hypothesis is that the sets of data have the same mean. (The way that ANOVA actually approaches this is to have a null hypothesis that the variation within groups is the same as variation between groups.)

An example In the illustrations of the use of the test for the packages I will be using the same example data set as in the *t*-test. With two samples of five observations each and each sample coming from a different cultivar of a crop plant.

The ideal ANOVA table that you would include in a write-up or publication should appear as something along the lines of the table shown here.

Source	d.f.	SS	MS	F	P
Cultivar	1	16.38	16.38	15.36	0.004**
Residual	8	8.53	1.07		
Total	9	24.92	2.77		

d.f. is degrees of freedom, SS is the sum of squares, MS is the mean square (sum of squares divided by degrees of freedom), *F* is the ratio of within-group variation to between-group variation (in this case it is the MS for cultivar/residual

MS). The *P*-value is the important one in this case it is less than 0.05 so we reject the null hypothesis that the two cultivars have the same mean grain size. The two asterisks indicate a highly significant result and are often used when $P<0.01$. Compare this table with the output obtained using the packages below to see where the various numbers have come from. (Note the word 'residual' often appears as 'error'.)

If you wish to write this result in the text of a report the standard way would be as follows: 'analysis of variance showed that the grain size of the two cultivars was significantly different ($F_{1,8}=15.36$, $P<0.01$)'. The two subscripted numbers after *F* indicate the degrees of freedom for the between-group and within-group variance.

SPSS As with the *t*-test all the data are placed into a single column and another column is used for the labelling of the groups (see Fig. 7.5). This may seem wasteful but it is a much easier system when it comes to more complicated analyses.

There are at least two routes to this test in SPSS. Unfortunately they give rather different outputs. I will describe them in detail below. I suggest you try both methods as the comparisons may help you understand how ANOVA tables work, and especially which parts of the tables to look at.

Method 1 Under the 'Analyze' menu choose 'Compare Means' and then 'One-way ANOVA …'. In the dialogue box that appears move 'Grain_sz' into the 'Dependent List:' box by first highlighting it and then clicking on the appropriate move button (blue arrow). Next move 'Cultivar' into the 'Factor' box. You can run the test now, although clicking 'Options…' allows you to request a 'Means plot' which is useful for visualization of the data, some 'Descriptive' statistics of the data and a test for 'Homogeneity of variance' (Fig. 7.6). Click 'Continue' to leave the 'Options' box, then 'OK' to run the test.

This is the minimum output you will get:

ANOVA

Grain size (mg)

	Sum of Squares	df	Mean Square	F	Sig.
Between Groups	16.384	1	16.384	15.362	.004
Within Groups	8.532	8	1.067		
Total	24.916	9			

This confirms that the data is 'Grain size (mg)' (or 'Grain_sz' if the variable has not been given an extra label). In the ANOVA table the top line is the important one; this is the variation between groups (i.e. between the two cultivars in this example). The second line is the variation within the groups that is being used as a comparison.

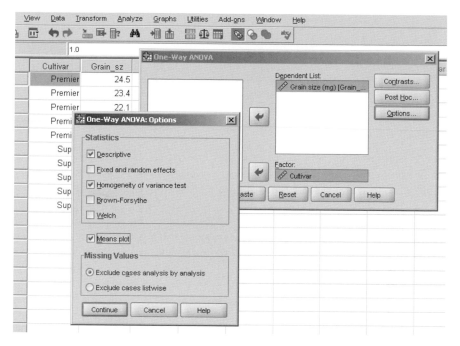

Fig. 7.6 Using SPSS for one-way ANOVA. The 'Options' dialogue box allows selection of useful additional output. The Welch or Brown–Forsythe options should only be selected if the Levene's test for homogeneity of variance gives a *P*-value less than 0.05.

The first column is 'Sum of Squares' (often called SS), then 'df' or degrees of freedom. As there are two groups in this example there is one degree of freedom between groups. There were five samples in each group giving two sets of four degrees of freedom and therefore eight in total. Next comes a calculation column: 'Mean Square' (often MS) is the sum of squares divided by the degrees of freedom. Both SS and MS are customarily included in ANOVA tables.

Finally comes the important bit; the *F*-ratio, here labelled 'F'. This is the mean square for between groups divided by that for within groups. If there is the same amount of variation between and within groups this will give an *F*-ratio of 1. In this case the *F*-ratio is over 15. SPSS gives you the *P*-value for this value of *F* with this degrees of freedom, labelling it as 'Sig.'. The value is 0.004, indicating that the two groups are highly significantly different, since *P* < 0.01.

'Options...': one of the assumptions of a basic ANOVA is that variances are equal. In the *t*-test in SPSS this is tested automatically. In ANOVA it is available under the 'Options...' button. Just check the box labelled 'Homogeneity-of-variance' and click 'Continue' before running the test. If you do this a little extra output appears before the ANOVA table.

→ **Oneway**

Test of Homogeneity of Variances

Grain size (mg)

Levene Statistic	df1	df2	Sig.
.762	1	8	.408

The important value is given the label 'Sig.'. The critical value is usually 0.05 and if the value given here is less than that ANOVA should not be used but a Mann–Whitney U test used instead. The value in the example is well above 0.05.

Method 2 Under the 'Analyze' menu choose 'General Linear Model' and then 'Univariate…'. In the dialogue box that appears move the variable with the observations ('Grain_sz' in the example) into the 'Dependent Variable:' box by first highlighting it and then clicking on the appropriate move button. Next move the variable with the group codes ('Cultivar' in the example) into the 'Fixed Factor(s):' box. Click 'OK' to run the test.

→ **Univariate Analysis of Variance**

Between-Subjects Factors

		Value Label	N
Cultivar	1	Premier	5
	2	Super	5

Tests of Between-Subjects Effects

Dependent Variable: Grain size (mg)

Source	Type III Sum of Squares	df	Mean Square	F	Sig.
Corrected Model	16.384[a]	1	16.384	15.362	.004
Intercept	6260.004	1	6260.004	5869.671	.000
Cultivar	16.384	1	16.384	15.362	.004
Error	8.532	8	1.067		
Total	6284.920	10			
Corrected Total	24.916	9			

a. R Squared = .658 (Adjusted R Squared = .615)

The first table confirms that the data have been grouped using 'Cultivar'; it gives the group labels ('Value Label') and the number of observations ('N'). Then it goes on to a rather less standard ANOVA table. Remember that this output is designed to cope with many factors and therefore extra lines, which appear totally superfluous here, are useful in more complex experiments. The important line is repeated twice, although you should read along the line labelled 'Cultivar'. This is the variation that was labelled 'Between Groups' in the first method. The line labelled 'Error' corresponds to the one labelled as 'Within Groups'.

The first column is 'Type III Sum of Squares' (don't worry about type III, just report it as SS in a table), then 'df' or degrees of freedom, and then 'Mean Square' (the mean square value is the sum of square value divided by the degrees of freedom). As there are two cultivars there is one degree of freedom. There were five samples within each cultivar giving two lots of four degrees of freedom and therefore eight in total for 'Error'.

Finally comes the important bit; the F-ratio, labelled 'F' here. This is the mean square for 'Cultivar' divided by that for 'Error'. The value of F is 15.362. SPSS gives you the P-value associated with this value of F and these degrees of freedom and labels it 'Sig.'. In biology we usually look for a value of less than 0.05. Here the probability is 0.004 which indicates that the mean grain sizes of the two cultivars are significantly different.

The output of this method is certainly not appropriate for inclusion in a write up as it stands: instead, create a table in the form I gave earlier in the example.

There are even more options to try in method 2. The 'Homogeneity tests' option is one that should usually be selected as it tests whether the data are appropriate for ANOVA. An output will appear similar to that in the data output on page 111. If 'Compare main effects' is selected with the variable defining groups ('Cultivar' here) in the 'Display Means for:' box, the following table will appear.

Pairwise Comparisons

Dependent Variable: Grain size (mg)

(I) Cultivar	(J) Cultivar	Mean Difference (I-J)	Std. Error	Sig.[a]	95% Confidence Interval for Difference[a]	
					Lower Bound	Upper Bound
Premier	Super	−2.560*	.653	.004	−4.066	−1.054
Super	Premier	2.560*	.653	.004	1.054	4.066

Based on estimated marginal means

*. The mean difference is significant at the .05 level.

a. Adjustment for multiple comparisons: Least Significant Difference (equivalent to no adjustments).

This is similar to the mean difference information given at the end of the t-test. It shows the differences between the mean values for the two factor levels (groups, or 'Cultivars' in the example), gives a standard error for the difference, a significance for the difference ('Sig.') and then 95% confidence intervals for the difference. If both 'Lower Bound' and 'Upper Bound' values have the same sign then the two factor levels are significantly different from each other.

R The observations should be in a single variable (vector) with another variable containing the group codes. Using the same data as the *t*-test example (page 104), the observations are in a variable called 'Grain' and the groups are coded in a variable called 'Cultivar'. A one-way ANOVA can be achieved in several ways in R: the simplest is to use the functions 'summary()' and 'aov()':

```
> summary(aov(Grain ~ Cultivar))

           Df  Sum Sq  Mean Sq  F value  Pr(>F)
Cultivar   1   16.384  16.384   0.15362  0.004422 **
Residuals  8   8.532   1.0665
---
Signif. codes: 0 '***' 0.001 '**' 0.01 '*' 0.05 '.' 0.1
' ' 1
```

This uses the R model syntax stating that 'Grain' is a function of 'Cultivar'. The output confirms the degrees of freedom, 'Df', gives the usual ANOVA calculation outputs of sum of squares, 'Sum Sq', and mean square, 'Mean Sq', which is the sum of squares divided by the degrees of freedom. Finally comes the important bit: the *F*-ratio, here called 'F value' which is the mean square of the source (here 'Cultivar') divided by the residual or error mean square. If there is no effect of the source the *F*-ratio with be 1. Here the value is 15.362. Finally the *P*-value is given, here labelled 'Pr(>F)', indicating that it is the probability of getting an *F*-ratio this large or larger if the null hypothesis is true. The *P*-value is well below 0.05 so we reject the null hypothesis and would report the result as '$F_{1,8}=15.36$, $P<0.01$', giving the effect and error degrees of freedom as a subscript. R adds asterisks to highlight significant results and gives a key. Here the two asterisks confirm that the *P*-value is between 0.001 and 0.01.

MINITAB As with the *t*-test, there are two ways of inputting the data. In the first the two groups should be in two columns; in the second all the data are in one column and another column is used to identify the groups. The second method appears wasteful but arranging the data in this way is required in more complex analyses.

Method 1 Input the data from the two groups into two separate columns and label appropriately. From the 'Stat' menu select 'ANOVA' then 'One-way (Unstacked)'. In the dialogue box ensure that both columns are in the 'Responses (in separate columns):' box. Click 'OK'.

(Or, if the command interface is enabled, type 'aovo c1 c2' at the MTB> prompt in the session window. Or you can input commands using 'Edit' menu then 'Command Line Editor'.)

You get this output:

```
One-way ANOVA: Premier, Super

Source   DF      SS     MS      F      P
Factor    1   16.38  16.38  15.36  0.004
Error     8    8.53   1.07
Total     9   24.92

S = 1.033    R-Sq = 65.76%   R-Sq(adj) = 61.48%


                                Individual 95% CIs For Mean Based on Pooled StDev
Level      N    Mean   StDev   -+----------+--------+---------+--------
Premier    5  23.740   1.218   (---------*-------)
Super      5  26.300   0.806                       (---------*-------)
                                -+----------+--------+---------+--------
                                22.8      24.0     25.2      26.4

Pooled StDev = 1.033
```

The first part of the output gives an ANOVA table in a fairly standard form (see the 'ideal' version in the example section above). The highly significant *P*-value of 0.004 means we reject the null hypothesis that the two cultivars have the same mean. However, it does not help us decide which group has the higher mean. The second section provides information about the two groups: number of observations ('N'), mean and standard deviation and then a rather primitive, but useful, graphical representation of the two group means and 95% confidence intervals of the means. In this example the confidence intervals of the two groups do not overlap (confirming that they are significantly different from each other) and clearly *Super* has the greater mean.

Method 2 Input all the observations into a single column. Use a separate column for coded labels for the two groups. In this example you would call the first column 'Grain sz' and the second 'Cultivar'. You must use integers as codes for groups.

From the 'Stat' menu select 'ANOVA' then 'One-way…'. Move the observed data into the 'Response:' box and the group codes into the 'Factor:' box. Don't worry about the comparisons button for only two groups, it will become useful when you have more than two groups. Click 'OK'

(Or type 'onew c1 c2' at the MTB> prompt in the session window. Or you can input commands using 'Edit' menu then 'Command Line Editor'.)

The output is identical to that from method 1 except that the two groups will be labelled with numbers rather than names.

Excel There is no direct method unless you have installed the Analysis ToolPak add-in.

1 Put the data into an Excel spreadsheet. This can be done in several ways: either exactly as in the example with one column for each cultivar and the names in the first row, or with all the data in a single column and an extra column having the names of the cultivars. (I will assume here that the data are in two columns, A and B, with labels in cells A1 and B1.)

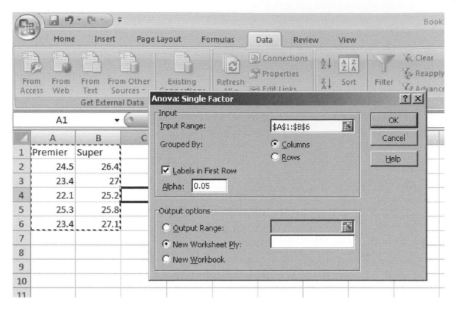

Fig. 7.7 One-way (single-factor) ANOVA using Excel. The area of the spreadsheet has been selected and is highlighted with a dashed line. Currently the output options will put the output onto a fresh sheet in the same workbook.

2 Select the 'Data' menu/ribbon and select 'Data analysis....', which should appear in a section of the ribbon called 'Analysis' (if this option does not appear you need to run Excel set-up and add the Analysis ToolPak add-in). Select 'Anova: Single Factor' and click 'OK'.

3 A dialogue box will appear. If the cursor is flashing in the 'Input range' box you can select the cells you wish to use for the analysis by clicking and dragging in the main sheet. If you select from cell A1 to the end of the data '$A$1:$B$6' should appear in the box. Alternatively you can just type in 'A1:B6'. The option 'columns' should be selected because the data are indeed in two columns representing different groups. The tick-box 'Labels in the first row' should be checked as the cultivar names are in the first row. Leave the 'Alpha' at 0.05 as this is the significance level which is chosen and $P<0.05$ is the usual level (Fig. 7.7).

4 At the bottom of the dialogue box is a section allowing you to determine where the output will appear. The default option is 'New Worksheet Ply' which means that the output will appear in a different sheet. If you want the output to appear on the same sheet as the data then you need to put a cell number in the 'Output range' box that will determine where the top left cell of the output will start.

5 Click on 'OK' and the following output will appear:

Anova: Single Factor						
SUMMARY						
Groups	*Count*	*Sum*	*Average*	*Variance*		
Premier	5	118.7	23.74	1.483		
Super	5	131.5	26.3	0.65		
ANOVA						
Source of Variation	*SS*	*df*	*MS*	*F*	*P-value*	*F crit*
Between Groups	16.384	1	16.384	15.3624	0.004422	5.317655
Within Groups	8.532	8	1.0665			
Total	24.916	9				

The first section gives summary information about the two groups of observations confirming the number of observations, sum, mean and variance.

The second section presents a rather standard ANOVA table: 'SS' is the sum of squares, 'df' degrees of freedom (there were 10 observations in all giving nine total degrees of freedom and there were two groups giving one degree of freedom for the 'between groups' variation), 'MS' is the mean square (=SS/df), F is the ratio of the variation between groups/within groups, 'P-value' is the important value as it shows the probability that the two cultivars have the same mean grain size. In this case $P=0.0044$ so we reject the null hypothesis that the two cultivars have the same mean grain size. The final value 'F-crit' is not usually quoted on an ANOVA table. It is the value of F required to achieve a $P=0.05$ with the degrees of freedom in this particular test. If alpha had been set at 0.01 in the ANOVA dialogue box then the 'F-crit' value would be 11.25863.

Mann–Whitney U

This test, also widely known as the Wilcoxon–Mann–Whitney test and less widely as the Wilcoxon rank sum W test, is the non-parametric equivalent of the independent samples *t*-test. It can only be used to test two groups. However, unlike the *t*-test and one-way ANOVA it does not make assumptions about homogeneity of variances or normal distributions. It is a typical 'rank' test, meaning that the raw data are converted into ranks before the test is carried out. The advantage of this is that it is ideal for situations where the highest value went off the scale or if extreme values are making the *t*-test undesirable.

The Mann–Whitney U test is less powerful than a *t*-test or one-way ANOVA, but you are less likely to find a significant result when there is no **real** difference. However, the lack of assumptions it makes about the distribution of the data make it the preferred test in many cases.

An example In the package illustrations for this test I use the same data set as for the t-test and one-way anova (page 104).

SPSS As with the *t*-test all the data are placed into a single column and another column is used for the labelling of the groups using integer code numbers that can be labelled as appropriate in the 'Variable View', 'Values' column. Under the 'Analyze' menu, choose 'Nonparametric Tests' and then select '2-Independent Samples...'. This will bring up a dialogue box for four different non-parametric tests. By default the 'Mann–Whitney U test' box should be selected. Put the variable with the observations ('Grain_sz' in the example) into the 'Test Variable List' box by selecting it and then moving it across. Then select the variable with the group codes ('Cultivar' in the example) as the 'Grouping Variable'. It will appear as 'Cultivar(??)'. You need to click the 'Define Groups...' button and then input '1' in 'Group 1' and '2' in 'Group 2' before clicking 'Continue'. This will return you to the first dialogue box. You will see that 'Cultivar(??)' is now 'Cultivar(1 2)'. Click 'OK' to run the test.

The output should appear as follows:

Mann-Whitney Test

Ranks

	Cultivar	N	Mean Rank	Sum of Ranks
Grain size (mg)	Premier	5	3.20	16.00
	Super	5	7.80	39.00
	Total	10		

Test Statistics[b]

	Grain size (mg)
Mann-Whitney U	1.000
Wilcoxon W	16.000
Z	−2.410
Asymp. Sig. (2-tailed)	.016
Exact Sig. [2*(1-tailed Sig.)]	.016[a]

a. Not corrected for ties.
b. Grouping Variable: Cultivar.

This confirms the test carried out. It also confirms that in the example the variable 'Grain size (mg)' was divided into two groups by the variable 'Cultivar' labelled as 'Premier' and 'Super'. Next comes some summary information. The two groups of 'Cultivar' each have five cases ('N'), giving 10 in total. The mean rank position of the two groups, with the smallest value given rank one and the largest rank ten, is also given. The 'Sum of Ranks' is given as it is an important part of the Mann–Whitney calculation, although it is not useful here. Even at this point it is clear that there is a difference between the two cultivars as the mean ranks are different.

The output from the test itself comes in a separate table labelled 'Test Statistics'. The first two rows are the output values from the Mann–Whitney and Wilcoxon versions of the test (U and W respectively). 'Z' is the statistic usually looked up in a table. Finally the rows labelled 'Asymp. Sig. (2-tailed)' and 'Exact Sig. [2*(1-tailed Sig.)]' give the crucial *P*-value. It is this value that you are interested in. Is it less than 0.05? In this example it is and we reject the null hypothesis that the two groups have the same median.

R The data to be compared should be in two variables. Here I have the example data in a dataframe called 't'. Just typing 't' at the command line will show the data:

```
> t
    Premier    Super
1   24.5       26.4
2   23.4       27.0
3   22.1       25.2
4   25.3       25.8
5   23.4       27.1
```

The Mann–Whitney test is also known as the Wilcoxon rank sum test and in R is reached through the 'wilcox.test()' function:

```
> wilcox.test(Premier,Super)

Wilcoxon rank sum test with continuity correction

data: Premier and Super
W=1, p-value=0.02118
alternative hypothesis: true location shift is not
equal to 0

Warning message:
In wilcox.test.default(Premier, Super) :
cannot compute exact p-value with ties
```

The output confirms the test used and that the default of a continuity correction has been applied (as is the default in MINITAB, but not SPSS). Next comes confirmation of the names of the two groups, the output of the test 'W' and the *P*-value, here below 0.05, so we reject the null hypothesis that the two sets of observations come from a distribution with the same median. Finally there is a description of the null hypothesis and then a warning that there are tied values in the data set (in the example 23.4 appears twice) and that means the data cannot be ranked in such a way to allow an exact version of the test.

To execute the test without the continuity correction (as happens by default in SPSS), this option has to be specified:

```
> wilcox.test(Premier,Super,correct=FALSE)

Wilcoxon rank sum test

data: Premier and Super
W=1, p-value=0.01597
```

I suggest that you use the default version with the continuity correction unless you need to compare results with a test where you know the correction has not been applied.

MINITAB Unlike the *t*-test there is only one way of inputting the data. Put the data into two columns, one for each group, and label accordingly. From the 'Stat' menu select 'Nonparametrics' then 'Mann Whitney...'. In the dialogue box move the two columns into the 'First sample:' and 'Second sample:' boxes (it doesn't matter which way round they are). Leave the other settings at their defaults or '95.0' and 'not equal'. Click 'OK'.

(Or, if the command interface is enabled, type 'Mann c1 c2' *at the MTB> prompt in the session window. Or you can input commands using* 'Edit' *menu then* 'Command Line Editor'.*)*

You get the following output:

```
MTB > Mann-Whitney 95.0 'Premier' 'Super';
SUBC> Alternative 0.
```

Mann-Whitney Test and CI: Premier, Super

```
           N    Median
Premier    5    23.400
Super      5    26.400


Point estimate for ETA1-ETA2 is -2.500
96.3 Percent CI for ETA1-ETA2 is (-4.301, -0.701)
W = 16.0
Test of ETA1 = ETA2 vs ETA1 not = ETA2 is significant at 0.0216
The test is significant at 0.0212 (adjusted for ties)
```

The output confirms the test carried out. Then it gives some summary information about the two groups: the number of observations ('N') and the median value. The next two lines give information about the difference between the two groups and quotes a range for the difference. Then comes a value for the test statistic 'W', given as 16.0 in the example.

The final two lines are the most important. They give the probability of the two medians being equal. The first probability ($P=0.0216$ in the example) does not account for tied data (two or more observations with exactly the same value) while the bottom line does. If there are several ties in the data then the second value will be higher and the second P-value for the test becomes more conservative.

In the example both versions give a $P<0.05$ so we reject the null hypothesis that the two groups have the same median value (i.e. in the example, the grain size for the two cultivars is different).

If the two significance results are either side of 0.05 then you will have to consider that a marginal result for the test suggests that you need to collect more data or measure with more precision to avoid tied data.

Excel There is no direct method even with the Analysis ToolPak installed.

Do the observations from more than two groups differ?

The groups can be either repeated (related) samples or they can be independent. Repeated measures are considered first.

Repeated measures

Repeated measures (a.k.a. related samples, matched samples) is an extension from paired data and occurs when a single individual or site is tested three or more times. A common example is in a 'before, during and after' design. Another possible use occurs when an individual is, or individuals of a clone are, divided and then subjected to three or more treatments. Two tests are considered below: the **Friedman test** and **repeated-measures** ANOVA.

Friedman test (for repeated measures)

The Friedman test is a non-parametric analogue of a two-way ANOVA. It makes no assumptions about the distribution of the data, only that they are measured on an ordinal scale. The Friedman is appropriate only if there is a single observation taken for each combination of factor levels. For repeated measures one of the factors must represent the repeat level, perhaps minutes, days or a measure of before, during and after a procedure. Then a the second grouping variable will be a standard factor such as region, species or treatment type. The null hypothesis is that observations in the same group (factor level) have the same median values. If the null hypothesis is rejected it shows that at least two groups have different medians, although it does not show which groups they are. Inspection will reveal which are the likely candidate groups (i.e. those with the highest and lowest medians).

It is the case for SPSS, R and MINITAB that the test has to be carried out twice if you wish to test both the conventional factor and whether the different sampling events have different median values.

The Friedman test is much less powerful than an equivalent ANOVA test if the data are normally distributed but makes fewer assumptions about the data so it is 'safer'.

An example The Friedman test could be used if the data comprise the number of cyanobacterial cells in 1 mm³ of water from six ponds, with samples taken on four different days and only one sample taken each day from each pond.

	Pond...					
Day	A	B	C	D	E	F
1	130	125	350	375	225	235
2	115	120	375	200	250	200
3	145	170	235	275	225	155
4	200	230	140	325	275	215

Note: there is only one observation for each pond/day combination.

SPSS Input the data using one column for each factor level (pond) in this example, arranged as in the table. Label the columns appropriately (letters A–F in the example). Make sure that each row corresponds to a level (group) of the repeated measure (e.g. if different times make sure that all the 'before' measures are in the first row, 'during' in the second and so on). However, this doesn't need a label at this point (i.e. you don't need a column labelled 'Day').

From the 'Analyze' menu select 'Nonparametric Tests' then 'K Related samples...'. In the dialogue box move all the columns containing data into the 'Test variables:' box. (Note that you can select the top item, hold shift and select the bottom item to select the whole list.) Make sure that the 'Friedman' box is checked (it should be selected by default). If you want any summary statistics about each of the groups you should click the 'Statistics...' button and check 'Descriptives' and click 'Continue' before you click 'OK'.

You should get the following output:

Friedman Test

Ranks

	Mean Rank
A	1.50
B	2.50
C	4.25
D	5.38
E	4.25
F	3.13

Test Statistics[a]

N	4
Chi-Square	11.259
df	5
Asymp. Sig.	.046

a. Friedman Test

The output confirms the test. Then it gives some rather unhelpful information about the mean rank of the observations in the different samples. The test output appears in the second table. First, the number of repeated or related samples is confirmed (four in the example). Then a test statistic is given followed by the degrees of freedom (one less than the number of groups being tested, so 5 d.f. from six ponds in the example). Finally the *P*-value is given, labelled as 'Asymp. Sig.'. If this value is less than 0.05 you reject the null hypothesis that the groups have the same median. In the example the *P*-value is 0.046. This value is just less than the critical 0.05 level. So we reject the null hypothesis that the ponds have the same median concentration of cyanobacteria.

To test another possible null hypothesis that the days all have the same median cell concentration would make the test a two-way analysis (see page 146).

R This is a simple test to execute in R using the function 'friedman.test()'. However, it is important to make sure that the data are arranged in a matrix. Here I show how the data can be input directly into R before the Friedman test is carried out:

```
> pondcells <-
+ matrix(c(130,125,350,375,225,235,
+ 115,120,375,200,250,200,
+ 145,170,235,275,225,155,
+ 200,230,140,325,275,215),
+ nrow=4,
+ byrow=TRUE,
+ dimnames=list(1:4,
+ c("A","B","C","D","E","F")))

> friedman.test(pondcells)

Friedman rank sum test
data: pondcells
Friedman chi-squared=11.259, df=5, p-value=0.04648
```

The output from the test first confirms the test used and the name of the data matrix. Then it gives the test statistic, here it's 11.259, then give the degrees of freedom 'df' and the *P*-value associated with that value of chi-square and degrees of freedom. In the example there are five degrees of freedom (number of ponds minus 1) and the *P* is marginally less than 0.05, so the null hypothesis that the ponds have the same median concentration of cyanobacterial cells is rejected, although with caution as the value is close to 0.05.

MINITAB Put the data into a single column. Then use the next two columns for integer codes of the two grouping variables. These codes must be integers but need not be consecutive.

From the 'Stat' menu select 'Nonparametric' then 'Friedman…'. In the dialogue box put the data in the 'Response' block, the conventional factor into the 'Treatment:' box and the repeated measure (perhaps a time, as in the example) into the 'Block:' box. Click 'OK'.

(Or, if the command interface is enabled, and assuming the observations are in column 1, the main factor in column 2 and the repeat factor in column 3, type 'Friedman c1 c2 c3' at the MTB> prompt in the session window.)

You get the following output:

```
Friedman Test: C1 versus Pond blocked by Day

S = 11.18 DF = 5 P = 0.048
S = 11.26 DF = 5 P = 0.046 (adjusted for ties)


Pond      N      Est Median     Sum of Ranks
1         4          150.52              6.0
2         4          166.35             10.0
3         4          295.10             17.0
4         4          293.85             21.5
5         4          240.10             17.0
6         4          192.19             12.5

Grand median = 223.02
```

Note: I coded the six ponds in the example with the numbers 1–6.

The output first confirms the test carried out. The first two rows give the results of the test, first with no correction for tied observations. The sample statistic for the test is 'S' and the 'DF.' is the degrees of freedom (one less than the number of factor levels) there were six ponds in the data set and therefore five degrees of freedom. Then there is a *P*-value associated with the 'S' result. In the example the *P*-value is just less than 0.05 so we reject the null hypothesis that the six ponds have the same median number of cells.

After the test results come some summary statistics about the groups giving number of observations ('N'), median and the total rank position of the observations in the whole data set ('Sum of Ranks'). The 'Grand median' is the median value from the whole data set.

By inspection you can determine that at least ponds 1 (A) and 4 (D) have significantly different medians as they are the extreme groups.

If you wish to determine whether the different sampling events have an effect on the median observations then you should repeat the test reversing the factors in the 'Treatment:' and 'Block' boxes. This makes the test into a 'two-way' analysis (see page 146).

(Or type 'Friedman c1 c3 c2' at the MTB> prompt in the session window.)

Excel There is no direct method even with the Analysis ToolPak installed.

Repeated-measures ANOVA

A two-way ANOVA may also be applied to this sampling design of only one measure for each combination of factor levels. The problem is that ANOVA makes the assumption that each of the factor levels is independent of all others. In a repeated-measures design, where a measurement is taken from an experiment on days 1, 2 and 3, the three days cannot be used as factor levels because measurements taken on day 2 are not independent of those already taken on day 1. The factor levels are also not equally dependent as day 1 is likely to be more similar to day 2 that it will be to day 3. There are ways around this problem and they usually entail a serious reduction in the degrees of freedom applied to the analysis.

ANOVA is a parametric test and therefore makes assumptions about the data. The data should be: continuous, normally distributed and with equal variances for the data in each factor combination (not the whole data set). In practice it is often impossible to test whether this is true because there are so few observations. Therefore it is usual to make a value judgement about whether the data are likely to conform to the assumptions.

In the example used above in the Friedman the data of number of cells per cubic millimetre are measured to the nearest 5 and is therefore discontinuous. However, there are more than 30 possible values so it would usually be acceptable to use ANOVA.

This test is only a special case of a two-way ANOVA where one of the factors defines the level of repeated sampling and the other the individuals or sampling sites. You could treat the analysis as a two-way ANOVA (usually without replication), see page 152 and below.

SPSS This test is reached from 'Analyze', then 'General Linear Model' then 'Repeated measures'. The package will not carry out the task using 'Repeated measures' if, as in the example, there is only one observation in each measure and only one individual ('pond' in the example) for each factor level.

However, if there are two or more individuals in each factor level, say if the ponds are grouped into 'ephemeral' and 'permanent', then the test can be carried out. Arrange the data so each measurement event is in a column (e.g. 'day') and each individual is on a separate row. There should be an additional column for the main effect (e.g. 'pondtype'). Use 'Analyze', then 'General Linear Model' then 'Repeated measures'. In the dialogue box insert the name of the repeated visit (e.g. 'day') and the number of times measured (four in the example) then click define.

In the next dialogue, move the measurement columns into the 'Within-Subjects Variables' box. There should be as many columns selected as you chose levels in the first window. Move the main effect to 'Between-Subjects Factor(s)'. Click 'OK'.

Masses of output is produced, reflecting the debates in the statistical literature about the most appropriate way to deal with repeated measures designs. At the bottom is an ANOVA style table. The row labelled with the name of the main effect in the first column ('Source') is the important one. Read the 'Sig.' value in the final column.

R In R, where the user specifies the model that's being used in the function, repeated measures designs can be achieved by specifying a different error from the default. Using the example where the dependent variable is 'Cells' and the independent variable (factor) is 'Pond' and the repeated visits are coded in 'Day':

```
> summary(aov(Cells~Pond*Day + Error(Cells/
(Pond*Day))))
```

MINITAB There is no repeated measure option in MINITAB. Although a two-way ANOVA (see page 152) can be carried out. In a repeated-measures design the F-ratio should be checked for significance with fewer degrees of freedom than for a conventional two-way ANOVA. This can be done by looking in statistical tables, or by using Excel as detailed below. MINITAB has its own look-up tables for many distributions. First put the number you want to look up into a 'constant' by typing 'Let K1 = 3.42', which puts the value 3.42 into the constant K1. Go to 'Calc' then 'Probability Distributions' then 'F...'. Select 'Probability density', put the appropriate main effect degrees of freedom in the 'Numerator degrees of freedom' and the error (or residual) degrees of freedom in the 'Denominator ...' the put K1, or wherever your F value was in the 'Input constant' box. In the example case 3.42 with five and 15 degrees of freedom is $P=0.029$, but with fewer degrees of freedom in the numerator to account for the repeated-measures design, 3.42 with two and 15 degrees of freedom is $P=0.041$.

[Or, if the command interface is enabled, type 'PDF 3.42;' at the MTB> prompt in the session window, then F 2 15 at the SUBC> prompt (giving the numerator and denominator degrees of freedom).]

Excel A two-way ANOVA (see page 152) can be carried out if the design is fully balanced with no missing data, although the F-ratio should be checked for significance with fewer degrees of freedom than for a conventional two-way ANOVA. This can be done with the function 'FDIST'. Use the 'Paste function' button, then 'Statistical', or type directly, 'FDIST(x,y,z)' where 'x' is the F-ratio reported in the ANOVA output, 'y' is the number of degrees of freedom for the main effect (five degrees of freedom from six ponds in the example) and 'z' is the number of degrees of freedom in the error. It is the value of 'y' (degrees of freedom of the main effect) that should be reduced.

Independent samples

This is the more usual type of analysis and occurs when a single individual or site is measured or tested only once. There will, therefore, be three or more totally separate groups of observations. Groups are often obvious; for example, pig, sheep, cow and horse. However, the distinction between the groups may be rather arbitrary, such as dividing samples by altitude bands or ranges of linear distance from a release point. Two tests are considered below: **one-way** ANOVA and the **Kruskal–Wallis test**.

If you get a significant result from one of these tests that is not the end of the story as it will not tell you which groups are different from which. Some kind of *post hoc* test (meaning 'after this') is required to allow you to interpret the results. Some of the many ***post hoc* tests** available are also described.

One-way ANOVA

The *t*-test is restricted to use with only two samples. When there are more than two groups you should use ANOVA. This is still a very simple use of ANOVA but it works well and is a test that every biologist should feel comfortable with.

ANOVA for three or more groups makes the same assumptions as for two groups: that the data are continuous, and the data within each group are at least approximately normally distributed and have equal variance. These assumptions should be tested before the test is carried out. The null hypothesis being tested is that each group has the same mean. (The way that ANOVA approaches this is to have a null hypothesis that the variation within groups is the same as variation between groups. The key is the ratio of the within-group variance and between-group variance that is termed the *F*-ratio.)

An example A researcher has grain-size data from three cultivars and wishes to determine whether there are any differences between them. The null hypothesis is that all three have the same mean (or that all three samples are taken from populations with the same mean). If there is a significant result (i.e. if $P < 0.05$) this indicates that at least one pair has different means; it does not tell you which pair. For convenience, in this example I will use the same example data set as in the *t*-test but with an additional third sample of five observations from a cultivar called Dupa which is coded as number 3. I would normally code the groups in alphabetical order. In this case as a third set of data has been added to a previously used set I have retained the labels 1 and 2 for those groups.

Cultivar name	Cultivar code	Grain size (mg)
Premier	1	24.5
	1	23.4
	1	22.1
	1	25.3
	1	23.4
Super	2	26.4
	2	27.0
	2	25.2
	2	25.8
	2	27.1
Dupa	3	25.5
	3	25.7
	3	26.8
	3	27.3
	3	26.0

If the data are analysed using ANOVA on a statistics package the output will not normally be directly suitable for presentation in a report. The ANOVA table that you would include in a write-up or publication should appear as something along the lines of the table below:

Source	d.f.	SS	MS	*F*	*P*
Cultivar	2	21.5	10.8	11.9	0.0014
Residual (error)	12	10.9	0.91		
Total	14	32.4			

Where *F* is the *F*-ratio (mean square of the main effect divided by the mean square of the error) and *P* is the *P*-value associated with the value of *F* and the degrees of freedom. Asterisks can be used to highlight significant levels of *P*. The best way to test that you are reading the output from your package correctly is to compare this table with the output obtained using the package.

SPSS As with many statistical tests, all the data are placed into a single column and another column is used for the labelling of the groups. This may appear to double the effort required but it is a much easier system when it comes to multiway analysis. *Post hoc* tests are important when trying to interpret the output from ANOVA. They are considered later.

As with the two-sample tests there is more than one route to this test in SPSS that gives the same results but with rather different outputs. I will consider two below in detail.

Method 1 Under the 'Analyze' menu choose 'Compare Means' and then 'One-Way ANOVA…'. In the dialogue box that appears move the variable with the observations ('Grain_sz' in the example) into the 'Dependent List:' box by first highlighting it and then clicking on the appropriate move button (blue arrow). Next move the variable with the group codes (in the example that is 'Cultivar') into the 'Factor:' box. (Ignore the 'Post Hoc' button for the moment, although you should always select a *post hoc* test when there are more than two groups. I will consider *post hoc* testing on page 138.) Click 'OK' to run the test.

The following will appear in the output window:

→ Oneway

ANOVA

Grain size (mg)

	Sum of Squares	df	Mean Square	F	Sig.
Between Groups	21.509	2	10.755	11.879	.001
Within Groups	10.864	12	.905		
Total	32.373	14			

This confirms the test as 'Oneway' (i.e. a one-way ANOVA), tells you that the observed data is 'Grain_sz' and then goes on to a standard ANOVA table. As is always the case in one-way ANOVA, the top line of the table is the important one: this is the variation between groups (i.e. between the three cultivars in this example). The second line is the variation within the groups that is being used as a comparison.

The first column is the 'Sum of Squares' (often just 'SS'), then comes 'df' or degrees of freedom. As there are three groups (cultivars) there are two degrees of freedom between groups (i.e. 3–1 = 2). There were five samples within each group giving three lots of four degrees of freedom and therefore 12 in total. Next come calculation columns 'Mean Square' (usually just 'MS'). The mean square value is the sum of squares value divided by the degrees of freedom.

Finally comes the important bit; the 'F' (or F-ratio). This is the mean square for between groups divided by that for within groups. If there is the same amount of variation between and within groups this will give an F-ratio of exactly 1. In this case the F-ratio is 11.879. This value could be looked up on an F table although SPSS gives you the P-value anyway, labelling it as 'Sig.'. The value is 0.001, indicating that *at least* two of the groups have means that are highly significantly different. There are three possible pairs with three groups: 1 and 2; 1 and 3; 2 and 3. It is the *post hoc* test that tells you which pairs are different from which, *not the ANOVA test itself.*

More complicated ANOVA tests are not possible under the 'Compare Means' menu so it may be better to become familiar with method 2.

'Options…': one of the assumptions of basic ANOVA is that variances are equal. In the *t*-test in SPSS this is tested automatically. In ANOVA it is available under the 'Options…' button. Just check the box labelled 'Homogeneity of variance test' and click 'Continue' before running the test. If you do this a little extra output appears after the ANOVA table.

Test of Homogeneity of Variances

Grain size (mg)

Levene Statistic	df1	df2	Sig.
.679	2	12	.526

The important value is given the label 'Sig.'. The critical value is usually 0.05 and if the value given here is less than that, ANOVA should not be used. A Kruskal–Wallis test should be considered instead (page 142).

Method 2 Under the 'Analyze' menu choose 'General Linear Model' and then 'Univariate…'. In the dialogue box that appears move the data column ('Grain_sz' in the example) into the 'Dependent Variable:' box by first highlighting it and then clicking on the appropriate move button. Next move the grouping variable ('Cultivar' in the example) into the 'Fixed Factor(s):' box. Either click 'OK' to run the test now or click on 'Options...' and select 'Homogeneity tests' first.

→ Univariate Analysis of Variance

Between-Subjects Factors

		Value Label	N
Cultivar	1	Premier	5
	2	Super	5
	3	Dupa	5

Levene's Test of Equality of Error Variances[a]

Dependent Variable:Grain size (mg)

F	df1	df2	Sig.
.679	2	12	.526

Tests the null hypothesis that the error variance of the dependent variable is equal across groups.
   a. Design: Intercept + Cultivar.

Tests of Between-Subjects Effects

Dependent Variable: Grain size (mg)

Source	Type III Sum of Squares	df	Mean Square	F	Sig.
Corrected Model	21.509[a]	2	10.755	11.879	.001
Intercept	9702.817	1	9702.817	10717.397	.000
Cultivar	21.509	2	10.755	11.879	.001
Error	10.864	12	.905		
Total	9735.190	15			
Corrected Total	32.373	14			

   a. R Squared = .664 (Adjusted R Squared = .608)

This output confirms the test used. Then gives a table containing only the number of observations for each factor level. The second table is the output requested in 'Options...' for 'Homogeneity tests'. A Levene's test has the null hypothesis that each factor level has the same variance. In the example the 'Sig.' for this test is well above 0.05, so there is no reason to reject this null hypothesis.

If the 'Sig.' is below 0.05 then an alternative test, such as the Kruskal–Wallis test should be considered (page 142).

Finally comes a rather less standard ANOVA table labelled as 'Tests of Between-Subjects Effects'. This output is designed to cope with many factors and therefore extra lines, that appear superfluous here, are useful in more complex experiments. The important line is repeated twice, although you should read along the line labelled 'Cultivar' (or whatever your main effect is labelled as). This is the test of the null hypothesis that between groups (factor levels) variation is the same as within group variation. The line labelled 'error' is the residual variation, or within group variation.

The second column is 'Type III, Sum of Squares' (don't worry about type III, just report it as SS in a table), then 'df' or degrees of freedom, then 'Mean Square' (the mean square value is the sum of square value divided by the degrees of freedom). As there are three cultivars there are two degrees of freedom. There were five samples within each cultivar giving three sets of four degrees of freedom and therefore twelve in total for 'Error'.

Finally comes the important bit; the F-ratio, labelled 'F' here. This is the mean square for 'Cultivar' divided by that for 'Error'. The value of 'F' is 11.879. SPSS gives you the P-value associated with this value of 'F' and these degrees of freedom and labels it 'Sig.'. We look for a value less than 0.05. Here the probability is 0.001 and indicates that the mean grain size of a least one pair of cultivars is significantly different. The ANOVA does not tell you which groups are different from which: a *post hoc* test is needed to determine that (page 138). (Note that SPSS does have a habit of reporting significance of 0.000, as is the case for the 'Intercept' in the example. This should always be reported as $P < 0.001$.)

The output table should be revised along the lines of the table I gave at the top of this section before it is included in a write-up or report.

You have some options at this point ('Options…'). For example, if you select the main effect and add it to 'Display Means for;' and check 'Compare main effects' you get output as follows:

Estimates

Dependent Variable: Grain size (mg)

Cultivar	Mean	Std. Error	95% Confidence Interval	
			Lower Bound	Upper Bound
Premier	23.740	.426	22.813	24.667
Super	26.300	.426	25.373	27.227
Dupa	26.260	.426	25.333	27.187

Pairwise Comparisons

Dependent Variable: Grain size (mg)

(I) Cultivar	(J) Cultivar	Mean Difference (I-J)	Std. Error	Sig.[a]	95% Confidence Interval for Difference[a]	
					Lower Bound	Upper Bound
Premier	Super	−2.560*	.602	.003	−4.227	−.893
	Dupa	−2.520*	.602	.004	−4.187	−.853
Super	Premier	2.560*	.602	.003	.893	4.227
	Dupa	.040	.602	1.000	−1.627	1.707
Dupa	Premier	2.520*	.602	.004	.853	4.187
	Super	−.040	.602	1.000	−1.707	1.627

Based on estimated marginal means
*. The mean difference is significant at the .05 level.
a. Adjustment for multiple comparisons: Sidak.

First a table gives the mean values for each level of the main effect (here with their group labels displayed). It also gives 95% confidence interval ranges. In the example it is clear that the groups Super and Dupa have very similar values, but that they are both different from Premier. The second table gives pairwise comparisons for each pair of factor levels. The differences in the mean differences and the 95% confidence intervals for the differences are also given. In this example I have selected the rather conservative Sidak method for pairwise comparisons (discussed below in the section on *post hoc* testing, page 138). In this example the differences between groups Premier and Super and between Premier and Dupa are significant ($P=0.003$ and 0.004 respectively), but the difference between groups Super and Dupa is not.

R There are several routes to this test in R. The one which provides the simplest output is using 'summary()' combined with 'aov()' with the tilde '~' to separate response variable (dependent variable, 'Grain' in the example) from the predictor variable (independent variable, or 'Cultivar' in the example). Here I have loaded the data from the example into R and attached it in a dataframe called 'oneway', before I run the test I confirm that the data are present and correct:

```
> oneway
     Grain  Cultivar
1    24.5   P
2    23.4   P
3    22.1   P
4    25.3   P
```

5	23.4	P
6	26.4	S
7	27.0	S
8	25.2	S
9	25.8	S
10	27.1	S
11	25.5	D
12	25.7	D
13	26.8	D
14	27.3	D
15	26.0	D

Here the factor levels were coded by letters. If the factor levels are coded by numbers R will treat the model 'Grain~Cultivar' as a request for a regression. The function 'as.factor()' is required to inform R that the numbers should be treated as labels for factor levels: 'Cultivar<-as.factor (Cultivar)'.

```
> summary(aov(Grain~Cultivar))
            Df    Sum Sq   Mean Sq    F value  Pr(>F)
Cultivar    2     21.509   10.7547    11.879   0.001428  **
Residuals  12     10.864    0.9053
---
Signif. codes:  0 '***' 0.001 '**' 0.01 '*' 0.05 '.' 0.1
 ' ' 1
```

Here there is no confirmation of the test used, just an ANOVA table laid out in a conventional way. The source of the variation, name of the predictor variable or independent variable or factor is given, then the degrees of freedom 'Df', then the sum of squares 'Sum Sq' and mean squares 'Mean Sq' (which is the sum of squares/degrees of freedom), then comes the F-ratio ('F value' here) which is the mean square of the factor/mean square of the residuals. Finally the P-value is given, here labelled 'Pr(>F)', indicating that the P is the probability of getting a value of F greater than or equal to this *if* the null hypothesis is true. In the example F is well above 1 and the P-value well below 0.01, so the result is highly significant. R provides an asterisk code of significance with '**' indicating a P-value of between 0.001 and 0.01.

MINITAB There are two routes to this test in MINITAB. You can either input all the data into a single column with a second column to code the groups (stacked) or you can put each group into a separate column (unstacked).

Method 1 Input all the data into a single column and use a second column to label the cultivars with integers (as in the example). Label the columns appropriately. From the 'Stat' menu select 'ANOVA' then 'One-Way…'. Move the observed data into the 'Response:' box and the group codes into the 'Factor:' box. The 'Comparisons…' button gives you access to the *post hoc* tests that are considered separately below. Click 'OK' to run the test.

[Or, if the command interface is enabled, type 'Oneway c1 c2' (assuming the data are in c1 and the group codes in c2) at the MTB> prompt in the session window. Or you can input commands using 'Edit' menu then 'Command Line Editor'.]

You get the following output:

```
One-way ANOVA: Grain sz versus Group

Source    DF      SS        MS        F        P
Group      2    21.509    10.755    11.88    0.001
Error     12    10.864     0.905
Total     14    32.373
S = 0.9515    R-Sq = 66.44%    R-Sq(adj) = 60.85%

                                  Individual 95% CIs For Mean
                                  Based on Pooled StDev
Level    N          Mean    StDev    +---------+---------+-------+--------
1        5        23.740    1.218    (----*----)
2        5        26.300    0.806                            (----*----)
3        5        26.260    0.764                            (----*----)
                                     +---------+---------+-------+--------
                                   22.8      24.0      25.2    26.4

Pooled StDev = 0.951
```

The first part of the output confirms the test carried out and gives an ANOVA table in standard form (see the 'ideal' version in the example section). The highly significant *P*-value (much less than 0.05) means we reject the null hypothesis that the cultivars have the same mean. However, it does not help us decide which groups are different from which. The second section provides information about the three groups: number of observations ('N'), mean and standard deviation and then a rather primitive graphical representation of the three group means and 95% confidence intervals of the means. In this example the confidence intervals of group ('Level') 1 does not overlap with groups 2 and 3 (confirming that they are significantly different from each other). The *post hoc* test will confirm this conclusion derived from inspection of the data (page 138).

Method 2 Input the data with a separate column for each of the groups (cultivars in the example) and label the columns appropriately. From the 'Stat' menu select 'ANOVA' then 'One-Way (unstacked) …'. In the dialogue box move all groups into the 'Responses (in separate columns):' box. Then click 'OK'. The output is identical to method 1 but with the column names appearing instead of the integer codes for the factor levels. A *post hoc* test can be accessed using the 'Comparisons' button.

Excel There is no direct method unless you have installed the Analysis ToolPak add-in. The method is essentially the same as for two groups.

1 Input the data onto an Excel spreadsheet. This can be done in several ways: either with one column for each cultivar and the names in the first row, or as in the example with the data from one cultivar in one row with an extra column having the names of the cultivars. (I will assume here that the data are in three columns A, B and C with names of the cultivars in cells A1, B1 and C1).

2 Select the 'Data' menu/ribbon and select 'Data analysis....' (if this option does not appear you need to the Excel options in the home menu and select 'Add-Ins' and add the 'Analysis ToolPak'). Select 'Anova: Single Factor'.

3 A dialogue box will appear. If the cursor is flashing in the 'Input range' box you can select the cells you wish to use for the analysis by clicking and dragging in the main sheet. If you select from cell A1 to the end of the data '$A$1:$C$6' should appear in the box. Alternatively you can just type in 'A1:C6'. The option 'Columns' should be selected because the data are indeed in two columns representing different groups. The tick-box 'Labels in First Row' should be checked as the cultivar names are in the first row. Leave the 'alpha' at 0.05 as this is the significance level which is chosen and $P < 0.05$ is the usual level used in biology.

4 At the bottom of the dialogue box is a section allowing you to determine where the output will appear. The default option is 'New Worksheet Ply' which means that the output will appear in a different sheet. If you want the output to appear on the same sheet as the data then you need to put a cell number in the 'Output range' box that will determine where the top left cell of the output will start.

5 Click on 'OK' and the following output will appear:

Anova: Single Factor

SUMMARY

Groups	Count	Sum	Average	Variance
Premier	5	118.7	23.74	1.483
Super	5	131.5	26.3	0.65
Dupa	5	131.3	26.26	0.583

ANOVA

Source of Variation	SS	df	MS	F	P-value	F crit
Between Groups	21.50933	2	10.75467	11.87923	0.001428	3.885294
Within Groups	10.864	12	0.905333			
Total	32.37333	14				

The usual components of an ANOVA table are all here: mean square and sum of square values, *F*-ratio and degrees of freedom. The important value is the

'*P*-value'. If this is less than 0.05 we reject the null hypothesis that all three cultivars have grains with the same mean size. In this case the value is well below that level at $P = 0.0014$ so we can say that the result is highly significant. However, this just gives the probability that two of the cultivars have different means: it does not show which ones. Inspection of the 'Summary' table shows that Super and Dupa have very similar means but Premier is different from them. There is no *post hoc* test available in Excel. If you need to have a *post hoc* test the easiest method it to carry out *t*-tests on each pair of factor levels in turn, but only when the *P*-value of the ANOVA is less than 0.05.

Post hoc *testing: after one-way* ANOVA

One of the commonest errors of omission I see is a one-way ANOVA carried out on, say, four groups with a *P*-value well under 0.05 and then no further investigation of which groups are different from which. It is important to realize that a significant result in the ANOVA will only show that at least one pair of the groups is significantly different. It does not identify which pair(s). When there are three groups that is only three possible pairs, with four groups that rises to six pairs and with five groups there are 10. *Post hoc* tests help you to make sense of this large number of possible comparisons by actually identifying which groups are significantly different from which.

The only problem is that the number of methods at your disposal for this task is overwhelming. For instance in SPSS you are offered a choice of seven different methods for answering the same question and these are only a subset of the number available.

The two tests I suggest you look for, and they will almost invariably give you the same results, are the least significant difference (LSD) test (a.k.a. Fisher's LSD test) and the Student Newman Keuls (SNK) test. The LSD test should *only* be carried out when the ANOVA result is significant. It uses the logic that if only significant results are examined there is no need to reduce the critical *P*-value (α) below 0.05 for the pairwise comparisons.

More conservative *post hoc* tests include the Bonferroni method, which is a general method for reducing the critical level required for significance below 0.05 achieved by dividing this value by the number of comparisons made. The Dunn–Sidák method is another general method of setting the critical value needed for significance (usually set to 0.05) when many pairwise comparisons are carried out at the same time.

It is conventional to show the results of a significant ANOVA by arranging the groups into ascending order of mean and then drawing lines under groups that are *not* significantly different. Therefore for the example I have been using here:

```
Premier        Dupa          Super
Cultivar 1     Cultivar 3    Cultivar 2
               _____
```

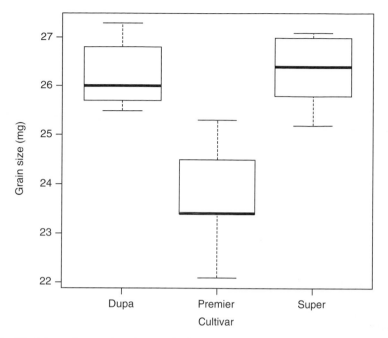

Fig. 7.8 A box-plot produced in R which can be used to visualize the results of a one-way ANOVA or a Kruskal–Wallis test.

MINITAB Before you click 'OK' to start a one-way test click the 'Comparisons…' button. This takes you to a dialogue box with four different tests. Select the 'Fisher's individual error rate:' option and make sure that the number in the box is a 5, this is the percentage significance level so 5 translates as critical P-value (α) of 0.05. This tests corresponds to the protected LSD test and should only be run if the P-value of the one-way ANOVA was less than 0.05.

When you run the test you get the following output after the standard ANOVA table:

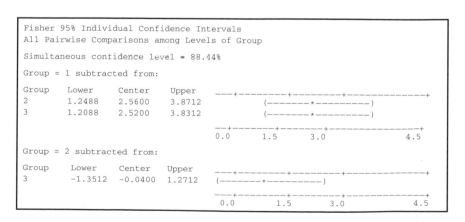

The graphical display of the mean and confidence interval for the means by groups shows quite clearly in this case that factor level '1' is very different from '2' and '3'. The 'Fisher's pairwise comparisons' section first confirms that 95% confidence intervals are being displayed. Finally comes a set of graphical representations of the pairwise comparisons between groups. With only three groups there will be two graphics, the first showing the differences between group 1 and 2, 1 and 3 and the second the difference between 2 and 3. With more factor levels there will be more output. In the example the difference between 1 and 2 and that between 1 and 3 is significant as the 95% confidence for the difference does not cross zero. However, for the difference between group 2 and 3 the 95% confidence for the difference does cross zero. In the example level 1 'Premier' is significantly lower than both level 2 ('Super') and level 3 ('Dupa') while level 2 is not significantly different from level 3. If i is the number of groups then there will be $i(i-1)/2$ comparisons (i.e. three for three groups, six for four groups, 45 for 10 groups) so the pairwise comparisons in this form become increasingly difficult to interpret.

Using the option 'Boxplots' of data after the 'Graphs…' button in the ANOVA dialogue will produce a useful graphical representation of the data organized by group. The option for 'Individual value plot' will plot every individual along with the mean values. This is fine when the data set is small, but will be too cluttered with a larger number of observations.

Excel No direct methods are available, although pairwise t-tests can be carried out *only* if the one-way ANOVA has already given a significant result, and this will give the same result as an LSD test. Be warned that if there are many factor levels there will be a lot of t-tests to be carried out.

Kruskal–Wallis test

This test is the non-parametric equivalent of the one-way ANOVA and has a null hypothesis that all samples are taken from populations with the same median. It can be used to test any number of groups. However, unlike one-way ANOVA it does not make assumptions about homogeneity of variances or normal distributions. It is a typical 'rank' test, meaning that the raw data are converted into ranks before the test is carried out. The advantage of this is that it is ideal for situations where the highest value went off the scale or if extreme values are present as these have a disproportionate influence on the results of parametric tests. This test *may* be used when there are only two samples, but the Mann–Whitney U test is more powerful for two samples and should be preferred.

This test is somewhat less powerful than one-way ANOVA, but you are less likely to find a significant result when there is no real difference (i.e. the probability of a type I error is decreased).

An example I will be using the same data set as for the one-way ANOVA so the results from the two tests may be compared.

SPSS Under the 'Analyze' menu, choose 'Nonparametric Tests' and then select 'K Independent Samples…'. This will bring up a dialogue box for three non-parametric tests. By default the Kruskal–Wallis test box should be selected. Put the variable containing the observations ('Grain_sz' in the example) into the 'Test Variable List:' box by selecting it and then moving it across. Then select the variable with the group codes ('Cultivar' in the example) as the 'Grouping Variable:'. It will appear as 'Cultivar(??)'. You need to click the 'Define Range…' button and then input '1' in 'Minimum' and '3' in 'Maximum' (or whatever the maximum value is in your data set) before clicking 'Continue'. This will return you to the first dialogue box. You will see that 'Cultivar(??)' is now 'Cultivar(1 3)'. Click 'OK' to run the test.

The output appears as follows:

→ NPar Tests

Kruskal-Wallis Test

Ranks

	Cultivar	N	Mean Rank
Grain size (mg)	Premier	5	3.20
	Super	5	10.40
	Dupa	5	10.40
	Total	15	

Test Statistics[a,b]

	Grain size (mg)
Chi-Square	8.655
df	2
Asymp. Sig.	.013

a. Kruskal Wallis Test
b. Grouping Variable: Cultivar

This confirms the test carried out is the Kruskal–Wallis. Next comes a table with confirmation that the variable 'Grain size' was divided into groups by the variable 'Cultivar'. Summary information about the three samples follows. In this example the groups of 'Cultivar' each have 'N' of 5 (cases or observations), giving 15 in total. The mean rank position of the groups (the smallest value of the 15 is given rank one and the largest rank 15) is also given. Even at this point it is clear that there is a difference between the cultivars as the mean rank of 'Premier' is very different to the other groups.

Finally comes the output from the test in a small table. The first row gives a 'Chi-square' value, degrees of freedom (there are three groups in the example, giving two degrees of freedom) and then a *P*-value, labelled as 'Asymp. Sig.'.

The Kruskal–Wallis test can make adjustments for tied observations (two observations with exactly the same value). The output given by SPSS does make some correction for ties. A data set with a large number of tied observations will be much less likely to give a significant result. See below for appropriate *post hoc* tests.

R The data should be in a single variable and the grouping labels in another variable. Using the data from the one way ANOVA example (page 129) the variables are named 'Grain' and 'Cultivar'. The 'kruskal.test()' is the required function in R:

```
> kruskal.test(Grain~Cultivar)

Kruskal-Wallis rank sum test

data: Grain by Cultivar

Kruskal-Wallis chi-squared=8.6555, df=2,
p-value=0.01320
```

This confirms the test being carried out and then the data that are being used. Here the dependent variable is 'Grain' and the independent or predictor variable is 'Cultivar'. The result of the test is given with the test statistic, the degrees of freedom ('df', being the number of groups minus 1) and then the *P*-value. In this case this value is well below 0.05 so we reject the null hypothesis that all three groups some from distributions with the same median. There is no *post hoc* test available, but see below.

MINITAB Input all the data into a single column and use a second column to label the cultivars with integers (as in the example). Label the columns appropriately. From the 'Stat' menu select 'Nonparametrics' then 'Kruskal–Wallis...'. In the dialogue box move the observed data into the 'Response:' box and the group codes into the 'Factor:' box. Click 'OK' to run the test.

[*Or, if the command interface is enabled, type* 'Krus c1 c2' *(assuming the data are in c1 and the group codes in c2) at the MTB> prompt in the session window. Or you can input commands using* 'Edit' *menu then* 'Command Line Editor'.]

You get the following output:

```
┌─────────────────────────────────────────────────────────────────────────┐
│  Kruskal-Wallis Test: Grain sz versus Cultivar                            │
│                                                                           │
│  Kruskal-Wallis Test on Grain sz                                          │
│                                                                           │
│  Cultivar          N      Median        Ave Rank               Z          │
│  1                 5       23.40              3.2           -2.94          │
│  2                 5       26.40             10.4            1.47          │
│  3                 5       26.00             10.4            1.47          │
│  Overall          15                         8.0                           │
│                                                                           │
│  H = 8.64    DF = 2    P = 0.013                                           │
│  H = 8.66    DF = 2    P = 0.013  (adjusted for ties)                      │
└─────────────────────────────────────────────────────────────────────────┘
```

The output confirms the test used. Then gives some summary information for each of the groups giving their integer codes. Number of observations is 'N' here, the median and mean rank are also given. The 'Z' value is used in the test calculation.

The last two lines give the test result. There are two versions depending on how they treat tied observations in the data. 'H' is a test statistic, 'DF' is the degrees of freedom (the number of groups minus 1). Finally the *P*-value, labelled 'P', is given. In the example it is far less than the critical 0.05 level and we reject the null hypothesis that the three groups have the same median. There is no *post hoc* test available to determine which group is different from which although this can be done by inspection of the summary statistics (and see below).

Excel   No direct method is available for the Kruskal–Wallis test in Excel, although it would be possible to carry out the test in the spreadsheet and obtain a significance value because the test statistic is distributed as chi-square.

Post hoc *testing: after the Kruskal–Wallis test*

The Kruskal–Wallis test has the same limitation as the one-way ANOVA in that a significant results just indicates that at least one pair of factor levels are significantly different from each other. The test does not indicate which pair, although inspection of the medians will show the extreme groups. However, there are two other pairwise combinations in a three-factor-level test and if there are four factor levels there is a total of six comparisons.

In one-way ANOVA a *post hoc* test will make all the pairwise comparisons and indicate which factor levels are different form which. There is not usually considered to be an equivalent for the Kruskal–Wallis test, but I suggest that, if the differences between each pair are an important thing to know, pairwise Mann–Whitney U tests (page 119) should be carried out. This method has exactly the same logic as the LSD test if it is *only* applied when the Kruskal–Wallis test gives a significant result.

There are two independent ways of classifying the data

If for each observation you have two factors (different ways of subdividing the data into groups) and the factors are independent of each other (i.e. there is no way the level of one factor can be deemed to be affected by the level of another) then there are several tests for analysing the null hypothesis that all factor levels have the same mean. For example you intend to record the beak widths of house sparrows in several different towns: your observations will be the measurement of beak width. One factor might be the sex of the bird and the other the town where it was collected.

In addition there may be a further null hypothesis that there is no interaction between the two factors under investigation. It is important to realize that interaction may only be investigated when there is more than one observation for each factor combination.

One observation for each factor combination (no replication)

There will be many circumstances where you wish to test the effect of two factors but are only able to take a very small number of observations. If you have only one observation for each factor combination then there are still tests you can perform to test the null hypothesis that each factor level has the same mean or median. The main difference in the interpretation of the results of these tests, when compared to tests with replication, is that there is no null hypothesis that there is no interaction between the two factors.

An example A trial of six different blends of fertilizers (coded with the letters U–Z) has been carried out on linseed crops on four different farms (coded 1–4). Factor 1 is the fertilizer and factor 2 the farm. On each farm six fields were used in the trial so it was only possible to use each of the six fertilizer blends once on each farm. The crop yields of the linseed (in kilograms per hectare) are given in the table.

	Fertilizer blend					
Farm	U	V	W	X	Y	Z
1	1130	1125	1350	1375	1225	1235
2	1115	1120	1375	1200	1250	1200
3	1145	1170	1235	1175	1225	1155
4	1200	1230	1140	1325	1275	1215

Friedman test

The Friedman test is a non-parametric analogue of a two-way ANOVA (page 152) that can only be used when there is a single observation for each factor combination (as in the data table above). The two null hypotheses are that the median values of each factor level are the same between columns and between rows. It does not make assumptions about the distribution of the data and can be used in any circumstance when the data are at least on an ordinal scale (can be put in a meaningful rank order). However, the test is rather conservative. Another problem is that in many packages the test has to be carried out twice: once to compare the rows and once for the columns.

SPSS The test compares columns only, so it has to be carried out twice. First you test the null hypothesis that there is no difference between the columns (fertilizer blends in the example). Input the data into SPSS in exactly the same format as it is in the table of raw data. You can either label the columns with the code letters U–Z or leave them as the default 'VAR00001' etc.

From the 'Analyze' menu choose 'Nonparametric Tests' and then 'K related samples...'. In the dialogue box that appears move all the columns into the 'test variables' box. Note, if the top item is selected and you hold shift while clicking on the bottom item the whole list is selected and can be moved to the 'Test Variables:' box in one go. Make sure that 'Friedman' is checked and click 'OK'.

This output should appear:

Friedman Test

Ranks

	Mean Rank
U	1.50
V	2.50
W	4.50
X	4.88
Y	4.50
Z	3.13

Test Statistics[a]

N	4
Chi-Square	10.396
df	5
Asymp. Sig.	.065

a. Friedman Test

The first section of the output confirms the test used then shows the mean rank position of the yield data for each of the six fertilizer blends (these are ranked 1 for the lowest on each farm and 6 for the highest). In the example the mean rank of blend 'U' is 1.5 on the four farms (it was lowest twice and second lowest twice).

The second section of the output ('Test Statistics') shows that there were four farms or cases ('N'), that the test output ('Chi-Square') was 10.396, there are 5 degrees of freedom 'df' and that the *P*-value ('Asymp. Sig.') is 0.065. This value is very close to the critical significance level of 0.05. So with such a conservative test we should seriously consider further trials.

You may wish to stop at this point or continue depending on whether you are interested in the differences between farms or not.

To test the null hypothesis that there is no difference between farms the data need to be transposed so that each farm is in a different column. This is very easy in SPSS: select 'Data' then 'Transpose…'. Move the column labels for each of the six fertilizer blends in the box labelled 'Variable(s):' and click 'OK'. If you had put a column in with the farm names that could be used at this point as a 'Name Variable:'. The data should be rearranged, possibly on a new data sheet, so the rows become columns. They are given the default names 'var001', 'var002' etc.

Carry out the Friedman test as above, make sure you don't use the variable 'CASE_LBL', and you get the following output:

Friedman Test

Ranks

	Mean Rank
var001	2.75
var002	2.17
var003	1.92
var004	3.17

Test Statistics[a]

N	6
Chi-Square	3.508
df	3
Asymp. Sig.	.320

a. Friedman Test

The details are as before. The most important number is the 'Asymp. Sig.' value of $P = 0.320$. We can therefore accept the null hypothesis that there is no difference between yields on different farms. In the example the variable names were left as the default SPSS names. These can be replaced easily by double-clicking on the column heading, or selecting the 'Variable View' and changing the 'Name' column.

R There are several ways of organizing the data, but it will need to be in a matrix format and then transposed in R. Assume that the data have been arranged as in the example with a column for each fertilizer blend and the data label at the top of the column. These data are then read into R. Here I have put the data into a text file labelled 'twowaym.txt' and imported it into a dataframe called 'frd':

```
> frd<-read.table("c:\\;temp\\;twowaym.txt", header=T)
> attach(frd)
> frd
      U       V       W       X       Y       Z
1     1130    1125    1350    1375    1225    1235
2     1115    1120    1375    1200    1250    1200
3     1145    1170    1235    1175    1225    1155
4     1200    1230    1140    1325    1275    1215
```

This data set is then put into a matrix called 'f'; notice the different notation in R when the data is in a matrix:

```
> f<-as.matrix(frd)
> f
        U       V       W       X       Y       Z
[1,]    1130    1125    1350    1375    1225    1235
[2,]    1115    1120    1375    1200    1250    1200
[3,]    1145    1170    1235    1175    1225    1155
[4,]    1200    1230    1140    1325    1275    1215
```

And a Friedman test is carried out on the matrix 'f':

```
> friedman.test(f)

Friedman rank sum test

data: f
Friedman chi-squared=10.3957, df=5, p-value=0.06477
```

This output confirms the test, gives the statistic and a value for degrees of freedom (one less than the number of columns) and gives a P-value. In this case the P-value is very close to 0.05. As the effect of fertilizer was the key interest in the example it might be sensible to stop the analysis now and set about collecting more data. However, if you want to look at the effect of the other factor ('farm' in the example) the data matrix needs to be transposed (so rows become columns). In R this is a very easy function called 't()'. Here I've put the transposed version of matrix 'f' into a new matrix 'g', visualized the new matrix to check it looks as I expect it to, then carried out another Friedman test on the transposed data:

```
> g<-t(f)
> g
      [,1]   [,2]   [,3]   [,4]
U     1130   1115   1145   1200
V     1125   1120   1170   1230
W     1350   1375   1235   1140
X     1375   1200   1175   1325
Y     1225   1250   1225   1275
Z     1235   1200   1155   1215
> friedman.test(g)

Friedman rank sum test

data: g
Friedman chi-squared=3.5085, df=3, p-value=0.3197
```

This time the *P*-value is well above 0.05, so we have no reason to reject the null hypothesis that data in the columns come from a distribution with the same median.

MINITAB Input the observations into a single column. Use two further columns for the two grouping variables (factors) replacing the labels for factor levels with integers. Label the columns appropriately.

From the 'Stat' menu select 'Nonparametrics' then 'Friedman…'. Put the observed data column into the 'Response:' box. Then put one of the factor columns into the 'Treatment:' box and the other into the 'Block:' box. If you are only really interested in one of the two factors then that should be in the 'Treatment:' box. Click 'OK' to run the test.

(Or, assuming the observations are in column 1, and the two factors in columns 2 and 3, type 'Friedman c1 c2 c3' *at the MTB> prompt in the session window. Or you can input commands using* 'Edit' *menu then* 'Command Line Editor'.)

You get the following output:

Friedman Test: Yields versus Fertilizer blocked by Farm

```
S = 10.32    DF = 5    P = 0.067
S = 10.40    DF = 5    P = 0.065  (adjusted for ties)

Fertilizer      N      Est Median      Sum of Ranks
1               4          1146.9               6.0
2               4          1161.5              10.0
3               4          1296.5              18.0
4               4          1255.2              19.5
5               4          1243.5              18.0
6               4          1192.7              12.5

Grand median = 1216.0
```

The output confirms the test carried out, then the name of the column containing the data with the factor being tested ('Fertilizer') in the example mentioned next. Then come two lines with the test statistic 'S', the degrees of freedom 'DF' (one fewer than the number of factor levels) and the P-value, 'P', associated with the test statistic. There are two versions of the test depending on how tied (equal) values are dealt with. The P-values will be similar unless there a lot of tied observations in the observed data. Use the P-value in the second row (adjusted for ties). In this case the P-value is close to the critical 0.05 level so that although we accept the null hypothesis that there is no difference in median yield at different fertilizer levels we would consider further investigation. After the test comes some summary information about the different factor levels giving number of observations ('N'), the estimated median yield for each factor level and then the mean rank position of the observations in each factor level.

This completes investigation of one factor. To investigate the other factor: go to the 'Stat' menu, select 'Nonparametrics' then 'Friedman…' and swap the column names in the 'Block:' and 'Treatment:' boxes.

(Or, if the command interface is enabled, type 'Friedman c1 c3 c2' *at the MTB> prompt.)*

For the example data this gives:

Friedman Test: Yields versus Farm blocked by Fertilizer

```
S = 3.45    DF = 3    P = 0.327
S = 3.51    DF = 3    P = 0.320  (adjusted for ties)

Farm        N              Est Median       Sum of Ranks
1           6                 1232.2                 16.5
2           6                 1213.4                 13.0
3           6                 1193.4                 11.5
4           6                 1259.7                 19.0

Grand median = 1224.7
```

Output is the same as for the first factor. In the example the P-value is well above 0.05 so we accept the null hypothesis that farms have the same median yield.

There is no way to determine the significance of any interaction between the two factors.

Excel There is no direct method for performing the Friedman test in Excel. It would be possible to calculate in this package but this test requires a lot of data sorting.

Two-way ANOVA *(without replication)*

It may seem slightly odd that a statistical test that relies on the comparison of variation can be used when there is only one observation for each factor level combination but the test is perfectly valid. The two grouping variables may be set by the experimenter (e.g. different concentrations) or 'naturally' occurring (e.g. different sites). The assumptions of the test are that the observed data are continuous, are approximately normally distributed, that the data would have about the same variance in each factor combination and that the grouping variables have at least two levels each that can be coded into integers. Obviously with only one observation in each factor combination there is no scope for testing the data against a normal distribution. Therefore it is up to the tester to use common sense or previous knowledge about the data under investigation.

An example The same example data set as for the Friedman test (see above) is used in the illustrations below. The data are crop yields that are rounded (therefore discontinuous) but there are clearly far more than 30 possible values so the assumption of continuous data can be accepted. The normal distribution might be more difficult as observations such as crop yields are often skewed to the right and would benefit from a log or square-root transformation. However, in this case we accept that the raw data are suitable for the ANOVA.

The results of the ANOVA should be presented in table form:

Factor	d.f.	SS	MS	F-ratio	P-value
Farm	3	11071	3690	0.797	0.515
Fertilizer blend	5	59763	11953	2.58	0.071
Error	15	69479	4631		

Where F is the important statistic (F-ratio = MS factor/MS error). Error is often referred to as residual.

Note that there are 23 degrees of freedom in total in this example. There were 24 observations in the total data set so that is correct: the two main effects have six and four cases to give five and three degrees of freedom, leaving 15 in the error. It is always sensible to check the number of degrees of freedom in the output to see if it matches your expectations.

There is no possibility of examining the interaction between the two factors as there is no replication. The interaction is an important part of the power of ANOVA where there is replication (see below).

In the example the P-value associated with the fertilizer blends was quite close to the critical 0.05 level. In a report this could be highlighted in the text as follows: 'There was no significant difference in yield between farms ($F_{3,15}=0.797$, $P>0.1$) but there was an indication that yields varied between fertilizer blends ($F_{5,15}=2.58$, $P=0.071$)'. The subscripted numbers quoted with the F-ratios are the degrees of freedom.

SPSS   Input the observed data into a single column. Use two separate columns to code the observations with integers to represent the factor levels. The integers do not have to be in sequence or start at 1. Label all the columns appropriately. In the example there will be a 'Yield' column, a 'Farm' column (containing the code numbers 1–4) and 'Fertiliz' column with number codes (1–6) replacing the letter codes from the original data. It is probably best to reduce the number of decimal places to zero at this point in the 'Variable View'. It is also possible to recode the numbers for the factor levels to more useful labels using the 'Values' column, although I have not done that here.

There are now at least two approaches to carrying out the test. They have different output so I will consider both. The first method assumes that both factors are 'fixed effects' (see page 193), the second considers 'random effects'. Fixed effects would be testing the differences between particular fertilizers and farms while random effects would be testing the differences between any farms and any fertilizers. In this current example it is likely that fertilizer would be a fixed effect, while farm would be a random effect although it will make no difference to the significance values in this scenario.

Method 1 From the 'Analyze' menu select 'General Linear Model' then 'Univariate...'. In the dialogue box move the observations ('Yield' in the example) into the 'Dependent Variable' box and the two factors into the 'Fixed Factor(s):' box. Before you carry out the test you must click on the 'Model...' button and bring up the Model box. Click on 'Custom', then select both of the factors in the 'Factors & Covariates:' list (hold down the Ctrl button while clicking on an item to add it to the current selection). From the drop down list under 'Build Term(s)' select 'Main effects' and click on the move arrow to move the factors to the 'Model' box. In the example this gives 'Fertiliz' and 'Farm' only in the 'Model' box. If the line 'Farm*Fertiliz' appears, then remove it. If you do not do this step, the analysis will not provide any significance values. Click 'Continue' and once back to the original dialogue box click 'OK' to run the test.

You will get the following output:

→ Univariate Analysis of Variance

Between-Subjects Factors

		N
Farm	1	6
	2	6
	3	6
	4	6
Fertiliz	1	4
	2	4
	3	4
	4	4
	5	4
	6	4

Tests of Between-Subjects Effects

Dependent Variable: Yield

Source	Type III Sum of Squares	df	Mean Square	F	Sig.
Corrected Model	70833.333[a]	8	8854.167	1.912	.133
Intercept	3.550E7	1	3.550E7	7664.673	.000
Farm	11070.833	3	3690.278	.797	.515
Fertiliz	59762.500	5	11952.500	2.580	.071
Error	69479.167	15	4631.944		
Total	3.564E7	24			
Corrected Total	140312.500	23			

a. R Squared = .505 (Adjusted R Squared = .241)

The output first confirms the test and then in a table 'Between-Subjects Factors' gives the number of observations in each level of the two factors. Then comes the ANOVA table itself, labelled as 'Tests of Between-Subjects Effects'. The name of the dependent variable (i.e. the data column) is confirmed. There is more information that you actually need in the table (compare this table to the 'ideal' output table given in the description of the example). The important lines are for the factor variables ('Farm' and 'Fertiliz' in the example) and the 'error' or residual. The 'df' column gives the degrees of freedom which is one less than the number of factor levels (in the example there were four farms so the factor 'Farm' has 3 d.f.). The (type III) sum of squares and mean square are given. Mean square is sum of squares/degrees of freedom. The last two columns give the important information. The 'F' column is the *F*-ratio (factor mean square/error mean square). The *P*-value is labelled 'Sig.' and given in the last column. In the example the *P*-value for 'Farm' is well above the critical 0.05 level so we accept the null hypothesis that all farms have the same mean yield. However, the *P*-value for 'Fertiliz' is 0.071, quite close to the critical level. We don't reject the null hypothesis but in any report we should note that the *P*-value is close to the critical level.

In this output it is best to totally ignore the lines labelled 'Corrected Model' (a combination of the two main effects), 'Total', 'Corrected Total' and 'Intercept' as not relevant to the two null hypotheses we are interested in.

Method 2 This time the test is carried out assuming that 'Farm' is a random factor and 'Fertiliz' is a fixed factor. From the 'Analyze' menu select 'General Linear Model' then 'Univariate…'. In the dialogue box move the observations into the 'Dependent Variable' box, 'Farm' in the 'Random Factor(s):' box and 'Fertiliz' into the 'Fixed Factor(s):' box. Before you carry out the test you must click on the 'Model…' button and bring up the Model box. Click on 'Custom', then select both of the factors in the 'Factors & Covariates' list. From the drop down list under 'Build Term(s)' select 'Main effects' and click on the move arrow to move the factors to the 'Model' box. In the example this gives 'Fertiliz' and 'Farm' only in the 'Model' box. If the line 'Farm*Fertiliz' appears, then remove it. Click 'Continue' and back in the main dialogue box you can either click 'OK' straight away or move first into the 'Options...' box where you can choose to display summary statistics for each of the variables. If you do this it can make interpretation of the output easier, although with no replication the distribution statistics cannot be calculated. Here I have selected 'Homogeneity tests'.

You get several tables of output. Here are the important ones:

Levene's Test of Equality of Error Variances[a]

Dependent Variable:
Yield

F	df1	df2	Sig.
	23	0	

Tests the null hypothesis that the error variance of the
dependent variable is equal across groups.
a. Design: Intercept + Fertiliz + Farm

Tests of Between-Subjects Effects

Dependent Variable:Yield

Source		Type III Sum of Squares	df	Mean Square	F	Sig.
Intercept	Hypothesis	3.550E7	1	3.550E7	9620.505	.000
	Error	11070.833	3	3690.278[a]		
Fertiliz	Hypothesis	59762.500	5	11952.500	2.580	.071
	Error	69479.167	15	4631.944[b]		
Farm	Hypothesis	11070.833	3	3690.278	.797	.515
	Error	69479.167	15	4631.944[b]		

a. MS(Farm)
b. MS(Error)

The first table, 'Levene's Test of Equality of Error Variances', would normally contain an estimate of the homogeneity of the variances in each factor combination. However, as there is only one observation in each, this cannot be calculated.

The second table, 'Tests of Between-Subjects Effects', contains the ANOVA information. The 'Dependent Variable' confirms the variable containing the data ('yield' in the example). The table is different to the 'ideal' table shown above as it has some extra lines. The important lines are those with the names of the two factors. The five columns of the table give the 'Type III sum of squares' (SS), 'df' (degrees of freedom), 'Mean Square' (MS), 'F' (*F*-ratio) and 'Sig.' (the *P*-value associated with the *F*-ratio and degrees of freedom). This is the value you compare with the critical level of 0.05. If the *P*-value is greater than 0.05 then you don't reject the null hypothesis that different levels of the factor have the same mean. In this example the *P*-value for 'Farm' is 0.515, well above 0.05, but that for 'Fertiliz' is, at 0.071, close to the critical level. This suggests that perhaps further investigation is required.

The line labelled as 'Intercept' should be ignored, although it might be noted that the 'Sig.' is given as '.000' when it should be reported as <0.001.

The lines labelled 'Error' give the error or residual line for a conventional ANOVA table, although the line used as error for the 'intercept' is actually the line for the random effect ('Farm' in this case). The error lines for the main effects should be reported in the ANOVA table.

R There are many ways of carrying out this test in R. I use the one that gives the easiest output to interpret. This uses the 'summary()' function applied to the analysis of variance function 'aov()'. Here I assume that the data have been loaded into R with the observations in one variable and the two grouping variables in two other columns. In the example I have imported this into a dataframe called 'two' and confirmed that I have labelled the columns:

```
> names(two)
[1] "Yield" "Blend" "Farm"
```

If the factor levels are coded as numbers it is important to tell R that the variable is a factor; otherwise, the variable may be used as a covariate. Use the function 'as.factor()' to do this: 'Blend<-as.factor(Blend)'.

The syntax for 'aov()' requires the response variable (your data) first, then the factors. If they are separated by an asterisk this will generate analysis of the variable and their interaction. However, in this unreplicated design there is a problem in that if the interaction is analysed there will be no degrees of freedom in the error (residual) which will result in the *F*-ratios being impossible to calculate:

```
> summary(aov(Yield~Farm*Blend))
            Df    Sum Sq   Mean Sq
Farm         3    11071     3690.3
Blend        5    59762    11952.5
Farm:Blend  15    69479     4631.9
```

Therefore in the unreplicated design R needs to be told to not use the interaction between the main effects. This is done by adding '-Farm:Blend':

```
> summary(aov(Yield~Farm*Blend-Farm:Blend))
            Df   Sum Sq   Mean Sq   F value   Pr(>F)
Farm         3    11071    3690.3    0.7967    0.51467
Blend        5    59762   11952.5    2.5804    0.07083.
Residuals   15    69479    4631.9
---
Signif. codes:  0 '***' 0.001 '**' 0.01 '*' 0.05 '.' 0.1
' ' 1
```

We now have the analysis we require. The table is in the familiar ANOVA style with degrees of freedom ('Df'), sum of squares, mean square and *F*-ratio ('F value') and then the *P*-value ('Pr(>F)'). In the example the effect of 'Farm' is well above 0.05, so we accept the null hypothesis, while the effect of 'Blend' is close to 0.05, so as we have a small data set we might consider further data collection. R provides a significance key at the end of the output. Here the row for 'Blend' is marked with a '.' which means the *P*-value lies between 0.05 and 0.1.

MINITAB There are at least two ways of carrying out this test in MINITAB. I will only discuss one here as it gives the most useful output.

Input all the data into a single column and then use the next two columns for the factors replacing factor level names with integer codes. Label the columns appropriately. From the 'Stat' menu choose 'ANOVA' then 'General linear model...'. Move the data column into the 'Responses:' box and the two factor columns into the 'Model:' box. Click 'OK' to run the test.

(Or, if the command interface is enabled, type 'GLM c1 = c2 c3' at the MTB> prompt in the session window. Or you can input commands using 'Edit' menu then 'Command Line Editor'.)

If you used the example you get the following output:

```
General Linear Model: Yields versus Fertilizer, Farm

Factor       Type      Levels    Values
Fertilizer   fixed        6      1, 2, 3, 4, 5, 6
Farm         fixed        4      1, 2, 3, 4

Analysis of Variance for Yields, using Adjusted SS for Tests

Source       DF    Seq SS    Adj SS    Adj MS      F      P
Fertilizer    5     59763     59763     11953    2.58   0.071
Farm          3     11071     11071      3690    0.80   0.515
Error        15     69479     69479      4632
Total        23    140313

S = 68.0584   R-Sq = 50.48%   R-Sq(adj) = 24.07%

Unusual Observations for Yields

Obs      Yields        Fit    SE Fit    Residual    St Resid
12      1140.00    1289.58     41.68     -149.58       -2.78 R

R denotes an observation with a large standardized residual.
```

The output confirms that you have carried out a 'General linear model' a general term for a family of statistical tests that includes ANOVA. Then it confirms the factors that have been used, how many levels there are for each factor and then the code numbers used for the factor levels (it pays to check these in case you made a mistake typing in the numbers!). Then comes the ANOVA table itself, stating the name of the observation variable (Yield in the example). The table has the usual columns except that it adds an additional SS column (with exactly the same numbers in as the usual SS column). 'Source' refers to the source of variation, 'DF' is degrees of freedom, 'SS' is sum of squares, 'MS' is mean square, 'F' is the F-ratio and 'P' the P-value. Compare the output to the ideal table I used in the description of the example. In the example the P-value for 'Farm' is well above 0.05 so we accept the null hypothesis but the P-value for is close to 0.05 so we may consider further investigation.

   Finally comes a list of unusual observations, these are rows in the data set that have a large residual value. 'Obs.' tells you which row number is unusual. These can often be typing errors so you should check the list against the original data set.

Excel There is no direct method unless you have installed the Analysis ToolPak add-in.
1 Put the data into an Excel spreadsheet in exactly the format it is in the example (see above) with the appropriate labels or code letters as column and row labels.
2 Select the 'Data' menu/ribbon and select 'Data Analysis....' (if this option does not appear you need to go to Excel options in the Home menu, select 'Add Ins' and add the 'Analysis ToolPak' add-in). Select 'Anova: Two Factor Without Replication'.
3 A dialogue box will appear. If the cursor is flashing in the 'Input Range' box you can select the cells you wish to use for the analysis by clicking and dragging in the main sheet. If you select from cell A1 to the end of the data '$A$1:$G$5' should appear in the box. Alternatively you can just type in 'A1:G5'. The tick-box 'Labels' should be checked as these are included. Leave the 'Alpha' at 0.05 as this is the significance level which is chosen and $P<0.05$ is the usual level.
4 At the bottom of the dialogue box is a section allowing you to determine where the output will appear. The default option is 'New Worksheet Ply' which means that the output will appear in a different sheet. If you want the output to appear on the same sheet as the data then you need to put a cell number in the 'Output range' box that will determine where the top left cell of the output will start.
5 Click on 'OK' and the following output will appear.

Anova: Two-Factor Without Replication

SUMMARY	Count	Sum	Average	Variance
1	6	7440	1240	11180
2	6	7260	1210	9230
3	6	7105	1184.167	1384.167
4	6	7385	1230.833	4054.167
U	4	4590	1147.5	1375
V	4	4645	1161.25	2606.25
W	4	5100	1275	11816.67
X	4	5075	1268.75	9322.917
Y	4	4975	1243.75	572.9167
Z	4	4805	1201.25	1156.25

ANOVA

Source of Variation	SS	df	MS	F	P-value	F crit
Rows	11070.83	3	3690.278	0.796702	0.51467	3.287382
Columns	59762.5	5	11952.5	2.58045	0.070826	2.901295
Error	69479.17	15	4631.944			
Total	140312.5	23				

The first section summarizes the data for each factor level of the two factors in turn. 'Count' is the number of observations and 'Average' the mean.

In the second section of the output there is a conventional ANOVA table. The 'Source of Variation' refers to the two factors. 'Rows' are farms and 'Columns' are fertilizer blends in this example. 'SS' is the sum of squares, 'df' is the degrees of freedom (there were 24 observations in all giving 23 total degrees of freedom; there were four farms giving 3 d.f. and six fertilizer blends giving 5 d.f.), 'MS' is the mean square (= SS/df) and F is the ratio of the factor MS/error MS. 'P-value' is the important value as it shows the probability that all factor levels (farms or fertilizer blends) have the same mean yield. In this case $P=0.51467$ for 'Rows' (i.e. farms) and $P=0.070826$ for 'Columns' (i.e. fertilizer) so we accept the null hypothesis that farms have the same mean yields and also accept the null hypothesis that the fertilizer blends have the same mean yield. However, as the P-value was quite close to 0.05 we may consider another trial.

The final values 'F crit' are not usually quoted on an ANOVA table. They are the values of F required to achieve a $P=0.05$ with the degrees of freedom in this particular test.

More than one observation for each factor combination (with replication)

The situation where you have two factors and more than one observation for each combination of factor levels is a very common one in biology. If there are two factors (ways of dividing the data into classes) then there are three hypotheses associated with the test: (1) that all levels of the first factor have the same mean, (2) that all levels of the second factor have the same mean and (3) that there is no interaction between the two factors. If you have no interest in the third null hypothesis then you might consider two separate one-way analyses. However, the extra power of the test that comes from investigation of the interaction is very great. Two tests are considered here: two-way ANOVA and the much less powerful non-parametric equivalent the Scheirer–Ray–Hare test.

Interaction

This concept warrants separate consideration. If the test gives a significant result for the interaction term it shows that the effects of the two factors in the test are not additive, which means that groups of observations assigned to levels of factor 1 do not respond in the same way to those assigned to factor 2. For example if you are measuring spiders from two locations you could have 'sex' as one factor and 'location' as the other (each spider can be assigned to one combination of sex and location). An ANOVA gives a significant result for the factor 'sex' and for the factor 'location' as well as the interaction term. This means that in some way the two sexes are responding differently in the two locations. The best way to interpret interaction is to plot out the means of the factor combinations roughly and inspect the graphs (Fig. 7.9).

SPSS Interactions can be visualized in SPSS fairly simply. Go to 'General Linear Model', 'Univariate…' and select the two factors and dependent variable in the usual way. Then click on the 'Plots…' button. In the 'Univariate: Profile Plots' dialogue box move one of the factors into the 'Separate Lines:' area and the other to the 'Horizontal Axis:' area. Then click on the 'Add' button. If there are no 'Factors:' available in the dialogue box they must be selected first in the main 'Univariate' dialogue box. By default the lines are different colours and different patterns might be more useful. The default lines can be changed: go to the 'Edit' menu and then 'Options…'. Select the 'Charts' tab and change the 'Style Cycle Preference' to 'Cycle through patterns only'. The exact sequence of patterns or markers that are used can also be altered in the 'Style Cycles' area (see Fig. 7.10).

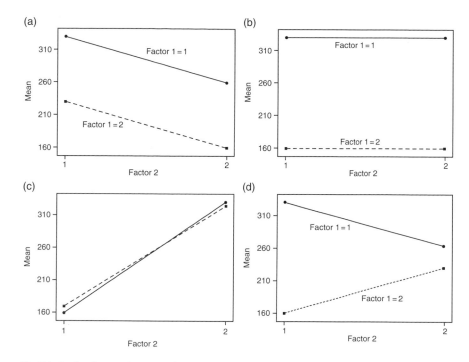

Fig. 7.9 In this figure there are four possible representations of the results of plotting mean values by factors after an analysis of variance. There are two factors labelled 'factor 1' and 'factor 2'. Both have two levels labelled '1' and '2'. Graph (a) shows a typical plot where both the main factors are significant but there is no interaction (lines are parallel). In (b) factor 1 is significant, factor 2 is not and there is no interaction. In (c) factor 2 is significant, factor 1 is not and there is no interaction. In (d) both factors are significant and there is also a significant interaction term: the lines are *not parallel* (i.e. the effect of factor 2 is different for the two groups of factor 1).

R   There is a separate plot function in R specifically for visualizing the interaction between factors: 'interaction.plot()'. Send the independent variable (factor) you want as the *x*-axis first, then the factor you want defined by separate lines and finally the dependent variable (response variable). The syntax for the unreplicated two-way ANOVA above would be either:

```
> interaction.plot(Farm,Blend,Yield)
```

or

```
> interaction.plot(Blend,Farm,Yield)
```

depending on whether you wanted 'Blend' or 'Farm' as the horizontal axis.

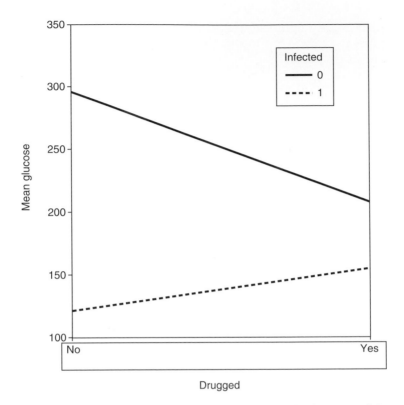

Fig. 7.10 In this example, produced in SPSS, the two lines are clearly not parallel, indicating that there is an interaction between the two factors. Here one factor is having an opposite effect on the two groups defined by the other factor. In the example the effect of the drug depends on whether the individual is infected or not.

MINITAB There is a very simple way to visualize interactions in MINITAB. Once the analysis has been carried out following the instructions below then a plot of the means for each factor and the effect of other factors can be produced. From the 'Stat' menu select 'ANOVA' and then 'Interactions plot...' (Fig. 7.11). Move the column with the data into the 'Response:' box and the grouping variables into the 'Factors:' box. You can carry out this procedure either before or after an ANOVA.

Excel Arrange the mean values for each factor combination in a table. Use row and column labels. For example:

		Subject unwell	
		0	1
Treated	0	295.66	207.50
	1	121.33	154.66

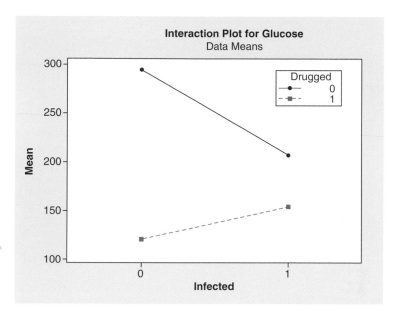

Fig. 7.11 This example uses the default 'interaction plot' option in MINITAB. There were two levels of two factors ('unwell' coded 0 or 1 and 'treated' coded 0 or 1). If there were no interaction between the two factors then the lines joining the means would be parallel. In this example the lines are not parallel and the treatment is clearly having an opposite effect on subjects to the control (i.e. when 'unwell' is 0 there is a decrease in mean level with treatment whereas when 'unwell' is 1 there is an increase).

Select these cells. Then from the 'Insert' menu/ribbon select within the 'Chart' area of the ribbon 'Line' and then the first '2-D line option'. A chart something like that shown in Fig. 7.12 will appear, although the default will be for coloured lines.

Two-way ANOVA (with replication)

This is a very powerful statistical test with three null hypotheses: the factor levels from the first main effect have the same mean, the factor levels from the second main effect have the same mean and the two main effect don't interact (described above). It is suitable when there are two independent ways of assigning the observations into groups and there is more than one observation per factor combination. Assumptions are the same as other ANOVA tests: the data are continuous (with more than 30 possible values), at least approximately normally distributed and the variation is the same in each factor combination. The assumptions appear restrictive but two-way ANOVA is not very sensitive to slight violations of the assumptions. If you have data that clearly don't fit the assumptions you have five choices: don't do the test at all, try to transform the data to make it fit the assumptions, carry out a Scheirer–Ray–Hare test *in lieu*, use two

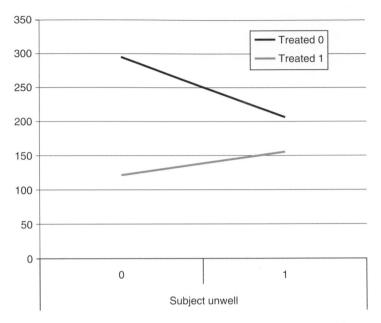

Fig. 7.12 An interaction visualized in Excel. The means have to be calculated first and then plotted. The lines are clearly not parallel, indicating that there in an interaction between the two factors. In this example when the subject is well ('unwell' is 0) there is a decrease in mean level with treatment but when 'unwell' is 1 there is an increase in the mean.

separate one way analyses instead or use the test but be very cautious if the *P*-values are anywhere near the critical 0.05 level.

Note: 'balanced designs' have equal numbers of observations in each factor combination. The statistical calculations are simpler for balanced designs and the tests more powerful. Therefore when you design a controlled experiment it is always sensible to plan for equal numbers of observations from each factor combination (of course this is not always possible when making field collections or when plants die in laboratory experiments). However, as you are not calculating the tests by hand I will not consider 'unbalanced' and 'balanced' designs separately. Make sure that you avoid factor combinations with no observations at all. Unfortunately Excel will only work with balanced designs for two-way analyses.

An example

A biologist is investigating the effect of light on food intake in starlings (*Sturnus vulgaris*). She sets up an experiment where birds are placed in individual aviaries of identical size with controlled lighting and are given an excess of food. Birds are sexed and four male and four female birds are exposed to either 16 h or 8 h of light (here termed 'long' or 'short' days). Each bird is monitored for 7 days and its total food intake in grams recorded. Each bird is only used once.

Day length	Sex							
	Female				Male			
Long (16 h)	78.1	75.5	76.3	81.2	69.5	72.1	73.2	71.1
Short (8 h)	82.4	80.9	83.0	88.2	72.3	73.3	70.0	72.9

There are 16 birds used in total. Four were exposed to each factor combination. Therefore, this is a balanced design.

The output from the statistical packages varies somewhat. It is usually not sensible to just put the whole output into a report. For a two-way ANOVA there is a standard way of displaying the output from the test. For the example here the output would be presented in the following way:

Factor	d.f.	SS	MS	F-ratio	P-value
Day length	1	42.3	42.3	8.00	0.015
Sex	1	316.8	316.8	60.00	<0.001
Interaction	1	27.0	27.0	5.12	0.043
Error	12	63.4	5.28		

SPSS Input all the observations in a single column (see Fig. 7.13). Use two further columns to input codes for the group labels associated with the two factors. The factor levels should be coded as integers. Label the columns appropriately (for the example, the factors have been labelled 'sex' and 'day_ len'). I suggest that the number of decimal places is adjusted in the 'Variable View' so the factors have no decimal places and the data has one. The 'Values' column of the 'Variable View' can be used to label the factor levels as 'male' and 'female', or whatever is appropriate.

From the 'Analyze' menu select 'General Linear Model' then 'Univariate…'. In the dialogue box move the observation column into the 'Dependent Variable:' box and the two coded factor columns into the 'Fixed Factor(s):' box. There is no need to use any of the options for a straightforward two-way design like this

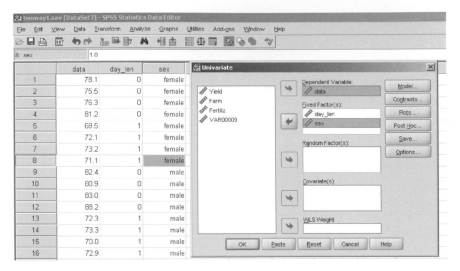

Fig. 7.13 Two-way ANOVA in SPSS. The two main effects have been defined as 'Fixed Factors' and the data have been moved to the 'Dependent Variable' box.

one, although selecting a 'Homogeneity of variance' test, as I have done in this example, might be wise. Click 'OK' to run the test.

Using the example you would get the following output:

Between-Subjects Factors

		Value Label	N
day_len	0		8
	1		8
sex	1	female	8
	2	male	8

Levene's Test of Equality of Error Variances[a]

Dependent Variable:data

F	df1	df2	Sig.
.888	3	12	.475

Tests the null hypothesis that the error variance of the dependent variable is equal across groups.

a. Design: Intercept + day_len + sex + day_len* sex

Tests of Between-Subjects Effects

Dependent Variable:data

Source	Type III Sum of Squares	df	Mean Square	F	Sig.
Corrected Model	386.130[a]	3	128.710	24.373	.000
Intercept	93025.000	1	93025.000	17615.591	.000
sex	316.840	1	316.840	59.998	.000
day_len	42.250	1	42.250	8.001	.015
day_len* sex	27.040	1	27.040	5.120	.043
Error	63.370	12	5.281		
Total	93474.500	16			
Corrected Total	449.500	15			

The output confirms the test used. Then gives the number of observations for each level of both factors. Individuals will be double-counted here (in the example there were 16 birds in total: eight female and eight male birds, eight in short days and eight in long). This is followed by a homogeneity of variance test. In this case there is no reason to reject the null hypothesis of equal variance.

Next comes the ANOVA table itself, labelled as 'Tests of Between-Subjects Effects'. The name of the dependent variable (i.e. the 'data' column) is confirmed. There is more information that you actually need in the table (compare this table to the 'ideal' output table given in the description of the example). The important lines are for the factor variables ('sex' and 'day_len' in the example), the interaction term ('day_len*sex' in the example) and the 'Error' or residual. The 'df' column gives the degrees of freedom which is one less than the number of factor levels (in the example there were two sexes and two day lengths so each has one degree of freedom). The degrees of freedom for the interaction term is the product of the degrees of freedom for the main effects (in this case both are one and 1×1 is 1). The (type III) sum of squares and mean square are given. Mean square is sum of squares/degrees of freedom. The last two columns give the important information. The 'F' column is the F-ratio (factor mean square/error mean square). The P-value is labelled 'Sig.' and given in the last column.

In this output it is best to totally ignore the lines labelled 'Corrected Model' (a combination of the two main effects), 'Total', 'Corrected Total' and 'Intercept' as not relevant to the three null hypotheses we are interested in.

In the example the P-value for 'sex' is given as '0.000'; this is a highly significant result but should always be reported as $P<0.001$. We can be sure that the two sexes have different food-intake levels although the test does not tell us which sex eats more: that must be done by inspection of the data. The P-value for the other factor ('day_len') is also less than 0.05 but the value of 0.015 indicates that the effect is not as strong as that for sex. Finally the interaction of the two factors also has a P-value less (albeit marginally) than 0.05, meaning that in this example the two sexes respond significantly differently to day length in the amount of food they eat. The direction of this effect can only be revealed by inspection of the mean values for each group (see the section on interaction above).

Some of the options that might prove useful for a two-way ANOVA include 'Descriptive statistics' and 'Homogeneity tests' in the options dialogue and a large range of *post hoc* tests (although with only two factor levels for each of the main effects a *post hoc* test just generates a warning message in SPSS). In the output below a 'Profile plot' has been generated in the 'Plots' dialogue with the two factors as 'Horizontal Axis' and 'Separate Lines'.

Descriptive Statistics

Dependent Variable:data

sex	day len	Mean	Std. Deviation	N
female	1	77.775	2.5290	4
	2	83.625	3.1753	4
	Total	80.700	4.1037	8
male	1	71.475	1.5714	4
	2	72.125	1.4751	4
	Total	71.800	1.4531	8
Total	1	74.625	3.8909	8
	2	77.875	6.5604	8
	Total	76.250	5.4742	16

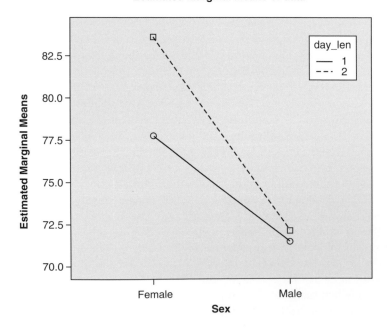

Estimated Marginal Means of data

The 'Descriptive Statistics' table usefully summarizes the full data set as well as each factor level and factor combination with means, standard deviations and number of observations. In the example I used 'Values' in the 'Variable View' to recode 'sex 1' as 'female' and 'sex 2' as 'male'. I didn't recode the factor levels in 'day_len' so they remain as '1' and '2'.

The 'Levene's Test of Equality of Error Variances' appears when the option 'Homogeneity tests' is selected.

R Make sure all the observations are in a single variable and that the two grouping variables (factors) are in two other variables. Import the data into R. There are many ways to achieve this test in R. A simple one which produces easily interpreted output is to use the 'summary()' function with the ANOVA function 'aov()'. Here I show the complete process, importing the data from a text file called 'starling.txt' into a dataframe I have called 'star'. This is then attached and visualized to check it is formatted correctly before the ANOVA is carried out:

```
> star<-read.table("c:\;\;temp\;\;starling.txt",
header=T)
> attach(star)
> star
       intake    sex        day
1      78.1      female     16
2      75.5      female     16
3      76.3      female     16
4      81.2      female     16
5      69.5      male       16
6      72.1      male       16
7      73.2      male       16
8      71.1      male       16
9      82.4      female     8
10     80.9      female     8
11     83.0      female     8
12     88.2      female     8
13     72.3      male       8
14     73.3      male       8
15     70.0      male       8
16     72.9      male       8
> summary(aov(intake~sex*day))
            Df   Sum Sq  Mean Sq  F value  Pr(>F)
sex          1   316.84  316.84   59.9981  5.224e-06  ***
day          1    42.25   42.25    8.0006  0.01522    *
sex:day      1    27.04   27.04    5.1204  0.04299    *
Residuals   12    63.37    5.28
```

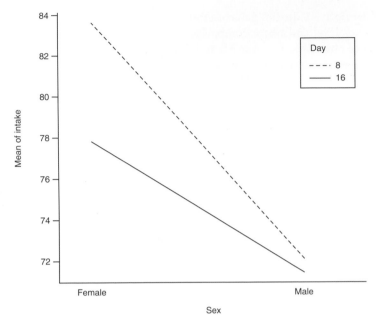

Fig. 7.14 An interaction visualized using R.

```
---
Signif. codes: 0 '***' 0.001 '**' 0.01 '*' 0.05 '.' 0.1
' ' 1
```

The ANOVA output is limited to the ANOVA table. It is in the usual format with a row for each source of variation: the two main effects and their interaction (here, 'sex', 'day' and 'sex:day'). For each source there is a degree of freedom 'Df'. The degrees of freedom for the interaction is reached by multiplication (here 1×1=1). The sum of squares and mean square are the usual calculation steps in an ANOVA table. When the degrees of freedom is 1 they will have the same value as mean square is sum of squares/degrees of freedom. Next comes the F-ratio ('F value') and finally the P-value ('Pr(>F)'). After the table there are asterisks to indicate the level of significance, with a key explaining the asterisks. In this case 'sex' is highly significant, P='5.224e–06', which might be better written as P=0.0000052, or simply P<0.001. It is marked with three asterisks, also indicating P<0.001. The effect of 'day' and the interaction of 'sex:day' are also significant with values of less than 0.05. This means that the two sexes are not responding to day length in the same way.

An interaction plot will visualize this relationship (see Fig. 7.14):

```
> interaction.plot(sex,day,intake)
```

MINITAB Input all the observations into a single column. Use separate columns to input the factor levels, coded as integers. Label the columns appropriately. There are then at least two ways of carrying out the test that give rather different output. I will consider both methods here.

Method 1 From the 'Stat' menu, select 'ANOVA' then 'General Linear Model…'. In the dialogue box put the column containing the data into the 'Responses:' box and then the two main effects (factors) into the 'Model:' box. To investigate interaction between the two factors the two column labels should be moved into the 'Model:' box again and an asterisk (*) put between them (in the example the 'Model:' box contained: sex 'day length' sex * 'day len' with day length appearing in quotes as the label is in two words). Click 'OK' to continue.

(*Or, if the command language has been enabled, type* 'glm c1 = c2 c3 c2*c3.' *at the MTB> prompt. Or you can input commands using* 'Edit' *menu then* 'Command Line Editor'.)

Using the example the following output appears in the 'Session' window:

General Linear Model: Uptake versus Day length, Sex

Factor	Type	Levels	Values
Day length	fixed	2	1, 2
Sex	fixed	2	0, 1

Analysis of Variance for Uptake, using Adjusted SS for Tests

Source	DF	Seq SS	Adj SS	Adj MS	F	P
Day length	1	42.25	42.25	42.25	8.00	0.015
Sex	1	316.84	316.84	316.84	60.00	0.000
Day length*Sex	1	27.04	27.04	27.04	5.12	0.043
Error	12	63.37	63.37	5.28		
Total	15	449.50				

S = 2.29801   R-Sq = 85.90%   R-Sq(adj) = 82.38%

Unusual Observations for Uptake

Obs	Uptake	Fit	SE Fit	Residual	St Resid
12	88.2000	83.6250	1.1490	4.5750	2.30 R

R denotes an observation with a large standardized residual.

First comes confirmation of the test (note that 'General Linear Model' includes ANOVA) and then the factors used, the number of levels (groups) and the integer codes assigned to the groups. (In the example there are obviously two levels for the factor 'sex' and they have been coded, alphabetically, as '1' for female and '2' for male).

The standard ANOVA table is next. It labels the factors and the interaction. The columns give degrees of freedom 'DF'; two versions of sums of squares 'Seq SS' and 'Adj SS'; the mean square 'Adj MS', which is the sum of squares divided by the degrees of freedom; the *F*-ratio 'F' which is the mean square divided by the error mean square and finally the *P*-value 'P'.

In the example the factor 'sex' has a *P*-value of 0.000 which should be reported as $P < 0.001$. We reject the null hypothesis that the two sexes have the same food intake confidently although we can't tell from the table which sex has the higher intake. The factor 'day length' has a *P*-value well below 0.05 so we also reject the null hypothesis that day length has no effect of food intake. The interaction has a *P*-value slightly less than 0.05 so we reject the null hypothesis that day length has the same effect on both sexes but note that the result is not highly significant.

The next part of the output gives a list of unusual observations: it is always wise to check that these have been input correctly into the package. This part of the output can be suppressed by going to the 'Results…' dialogue within 'General Linear Model' before running the test and selecting the option 'Analysis of variance table'.

Method 2 From the 'Stat' menu, select 'ANOVA' then 'Two way…'. In the 'Two-way Analysis of Variance' box that appears move the column with the data to the 'Response:' box and the two main effects or factors ('sex' and 'day length' in the example) to the 'Row factor:' and 'Column factor:' boxes. Click 'Display means' for extra useful information in the output. Click 'OK'.

(Or, if the command language has been enabled, type 'twoway c1 c2 c3;' *at the MTB> prompt and* 'means c2 c3.' *at the SUBC> prompt.)*

The following output appears in the 'Session' window:

```
Two-way ANOVA: Uptake versus Sex, Day length

Source          DF        SS          MS         F         P
Sex              1     316.84     316.840     60.00     0.000
Day length       1      42.25      42.250      8.00     0.015
Interaction      1      27.04      27.040      5.12     0.043
Error           12      63.37       5.281
Total           15     449.50

S = 2.298    R-Sq = 85.90%    R-Sq(adj) = 82.38%

                   Individual 95% CIs For Mean Based on Pooled StDev
Sex    Mean     +---------+---------+---------+---------+----
0      80.7                                      (----*----)
1      71.8       (-----*-----)
                 +---------+---------+---------+---------+----
                    70.0      73.5      77.0      80.5

                   Individual 95% CIs For Mean Based on
Day                Pooled StDev
length    Mean     ------+---------+---------+---------+---
1       77.875                       (-------*--------)
2       74.625       (--------*--------)
                   ------+---------+---------+---------+---
                       74.0      76.0      78.0      80.0
```

The ANOVA table is very much as it would appear in a publication or report with DF, SS, MS, F and P clearly laid out (see description of data output above for more details). The 'Two-Way…' route allows the differences between factor levels to be seen semi-graphically. In the example the 95% confidence intervals for food taken for the two groups are very clearly different, reflecting the very low P-value. The intakes differ by day length, but the 95% CIs overlap slightly, reflecting the much higher, though still significant, P-value. It is also clear which groups are higher with females (group 1) eating more than males (group 2) and those in short days (group 2) eating more than those in long days.

ExceL This test is only possible in Excel if the design is balanced (i.e. each factor combination, long/female, short/female, etc., has the same number of observations). The data must be input so that one of the factors is separated into different columns and the other in sets of rows. The columns must be labelled with group labels and the first row of each group likewise.
The example was set up in Excel like this:

	Long	Short
Female	78.1	82.4
	75.5	80.9
	76.3	83
	81.2	88.2
Male	69.5	72.3
	72.1	73.3
	73.2	70
	71.1	72.9

From the 'Data' menu/ribbon select 'Data Analysis' then 'ANOVA: two factor with replication' (the Data Analysis option is not there you have to add the Analysis ToolPak through the 'Excel Options' in the 'Home' menu). In the dialogue box that appears you must select the range of cells that contains not only the data but also the labels. This can be done either by dragging the mouse over the area of the relevant section of the spreadsheet or typing the top left and bottom right cell codes with a colon ':' between them (A1:C9 in this case). You must indicate how many observations there are in each factor combination (in this case four). Leave the 'Alpha:' as 0.05 (this determines the critical F-value which is quoted on the final table for comparison; i.e. the F-value required to reject the null hypothesis).
Finally, to select where the output will appear select one of the options in the bottom half of the box. Beware: the output covers a lot of cells. The output is rather extensive:

SUMMARY	Long	Short	Total				
Female							
Count	4	4	8				
Sum	311.1	334.5	645.6				
Average	77.775	83.625	80.7				
Variance	6.39583	10.0825	16.84				
Male							
Count	4	4	8				
Sum	285.9	288.5	574.4				
Average	71.475	72.125	71.8				
Variance	2.46917	2.17583	2.11143				
Total							
Count	8	8					
Sum	597	623					
Average	74.625	77.875					
Variance	15.1393	43.0393					
ANOVA							
Source of Variation	*SS*	*df*	*MS*	*F*	*P-value*	*F crit*	
Sample	316.84	1	316.84	59.9981	5.2E-06	4.74723	
Columns	42.25	1	42.25	8.00063	0.01522	4.74723	
Interaction	27.04	1	27.04	5.1204	0.04299	4.74723	
Within	63.37	12	5.28083				
Total	449.5	15					

The first part of the output confirms the test carried out and then presents a summary table giving number of observations, sum, mean and variance for each factor combination, factor level and total. Some of this information is useful, especially when trying to interpret a significant result from the interaction term. (Note that on some very old versions of Excel, the average and variance numbers given were incorrect for column and row totals).

Then comes a rather standard ANOVA table that will require minimum editing before it can be included in a report. If you compare it with the 'ideal' table in the example you see how similar it is. The two factors are, unfortunately, only assigned as 'columns' and 'sample' although inspection of the original data in Excel will quickly determine which factor is which. The residual or error line is labelled as 'within'. The columns of figures are sum of squares (SS), degrees of freedom (df), mean square (MS, where MS = SS/df), F-ratio (F, where $F = MS/MS_{error}$), the P-value and then the F-ratio that was required to reach a P–value of 0.05. The P-values are sometimes given in scientific notation so the '5.2E-06' in the example translates as 5.2×10^{-6} or 0.0000052; that is, extremely highly significant.

In the example all three P-values are less than 0.05 so all three original null hypotheses are rejected.

Scheirer–Ray–Hare test

It is quite easy (especially with fairly small data sets) to calculate a two-way ANOVA using ranks in an extension of the Kruskal–Wallis test called the *Scheirer–Ray–Hare test*. Non-parametric two-way ANOVA is usually deemed impossible in statistics books but the Scheirer–Ray–Hare test is a non-parametric equivalent of a two-way ANOVA with replication. It is based on ranks so is suitable for any situation where the data can be put into order. It makes few assumptions about the distribution of the data. However, the test is not widely available in statistical packages. This test is conservative and has much lower power than the parametric ANOVA. Unfortunately this is an area of considerable debate among statisticians and there are some workers who are unconvinced that the Kruskal–Wallis test can be extended to a two-way analysis in this way. I suggest that if you do use it you do so with some caution and perhaps consider a generalized linear model with an error structure that doesn't require normal errors.

An example

We will use the same data of starlings grouped by sex and day length as for the two-way ANOVA. Unlike the standard ANOVA which rejects all three null hypotheses, the Scheirer–Ray–Hare test accepts two of the three null hypotheses associated with the experiment: there is no interaction between the two factors and birds in different day lengths consume the same amount of food. This illustrates how much more conservative the Scheirer–Ray–Hare test is than ANOVA.

SPSS

1 Set out the data in exactly the same way as for parametric ANOVA. Sort the data (ascending) using the magnitude of the observations. From the 'Data' menu select 'Sort Cases…'. In the dialogue box move the data column name into the 'Sort by:' box and select 'Ascending' (although it does not matter whether you choose ascending or descending). Click 'OK' and the rows will be shuffled into rank order.

2 Next you should assign ranks to replace the actual data (assign 1 to the smallest etc.). You can either write over the original data observations or make a new column called 'rank'. Type the row number into this column. This is trivial when there are only 16 observations as in the example, but with hundreds of observations it would be better to import a list of ascending numbers from a spreadsheet.

3 Carry out a standard (parametric) ANOVA including the interaction. From the 'Analyze' menu select 'General Linear Model. then 'Univariate…'. In the dialogue box put the 'rank' column into the 'Dependent:' box and the two coded factor columns into the 'Fixed Factor(s)' box. Click 'OK' to run the test.

You will get this output if you use the example:

Between-Subjects Factors

		Value Label	N
sex	0	female	8
	1		8
day_len	1	male	8
	2		8

Tests of Between-Subjects Effects

Dependent Variable: rank

Source	Type III Sum of Squares	df	Mean Square	F	Sig.
Corrected Model	285.000ᵃ	3	95.000	20.727	.000
Intercept	1156.000	1	1156.000	252.218	.000
sex	256.000	1	256.000	55.855	.000
day_len	25.000	1	25.000	5.455	.038
sex* day_len	4.000	1	4.000	.873	.369
Error	55.000	12	4.583		
Total	1496.000	16			
Corrected Total	340.000	15			

a. R Squared = .838 (Adjusted R Squared = .798)

4 This is not the end point for this test. You have to do the next bit by hand. Calculate the mean square (MS) value for the 'Corrected Total' row. This is not done in the SPSS output. It is the value of the 'Type III Sum of Squares' divided by the degrees of freedom. Call this value MS_{total} (in the example MS_{total} is $340/15 = 22.667$).

The test statistic for the three null hypotheses is the value for relevant sum of squares (SS) divided by MS_{total}. Calculate SS/MS_{total} for each factor and the interaction.

Using two new columns in the SPSS spreadsheet, type in the values of SS/MS_{total} and degrees of freedom using the same degrees of freedom as in a conventional ANOVA (i.e. one less than the number of factor levels for the two main factors and the degrees of freedom for the main factors multiplied together for the interaction). Label the columns 'out' and 'df'.

Then, to get the appropriate P-values for these values, from the 'Transform' menu select 'Compute Variable...'. In the dialogue box that appears, type the name of the column you want the result to go to in the 'Target Variable' box (e.g. 'pvalue'). Then in the 'Numeric expression:' box type '1-' (i.e. one, minus) before selecting 'CDF & Noncentral CDF' from the 'Function group:' list and then 'Cdf.Chisq' from the 'Functions and Special Variables:' list. Move that into the 'Numeric Expression:' box. Finally, replace the first question mark with the name of the variable where you input the SS/MS$_{total}$ values and the second question mark with the name of the column with the degrees of freedom. If you used 'out' and 'df' as I suggested you should have '1-CDF.CHISQ(out.df)'. Click 'OK' and the new column 'pvalue' will appear on the right-hand side of the spreadsheet with P-values from a chi-square table appropriate to the SS/MS$_{total}$ values and degrees of freedom. (You may need to increase the number of decimal places above 2 in the 'Variable View'.) *Note*: you could look up the number in a chi-square table if you prefer.

So, for the example the Scheirer–Ray–Hare output could be displayed as:

	SS	SS/MS$_{total}$	d.f.	P-value
Day length (factor)	25	1.10	1	0.2943
Sex (factor)	256	11.29	1	0.0008
Day length*sex (interaction)	4	0.176	1	0.6748

The results suggest that the only null hypothesis that must be rejected is that both sexes consu me the same amount of food, as the P-value is less than 0.001.

R There are several routes to achieving this test in R. Here I'm assuming that the data have been imported as for the two-way ANOVA with replication. In my example that is a dataframe I have called 'star'.
1 make a new variable called 'myrank' that uses the 'rank()' function using 'intake' with 1 assigned to the smallest value, etc.

```
> myrank=rank(star$intake)
```

This new variable should then be attached to the dataframe which is called 'star' in this example:

```
> star$rank=myrank
> attach(star)
```

You can then visualize the contents of the first few rows of the dataframe to check that the new variable 'rank' has been added:

```
> star[1:5,]
    intake    sex       day   rank
1   78.1      female    16    11
2   75.5      female    16    9
3   76.3      female    16    10
4   81.2      female    16    13
5   69.5      male      16    1
```

2 Carry out an analysis of variance as in the two-way ANOVA with replication above, but using the variable 'rank' as the response variable:

```
> summary(aov(rank~sex*day))
            Df   Sum Sq   Mean Sq   F value   Pr(>F)
sex         1    256      256.000   55.8545   7.494e-06 ***
day         1    25       25.000    5.4545    0.03769 *
sex:day     1    4        4.000     0.8727    0.36862
Residuals   12   55       4.583
```

3 This isn't the end point. You have to calculate the mean square for the total row. This is not visible in this output, but is the sum of the 'Sum Sq' values (here $256+25+4+55=340$) divided by the total degrees of freedom (here it is 15), giving a value of $340/15=22.667$ in the example.

The test statistic for the Scheirer–Ray–Hare test is the value for the sum of squares divided by the MS_{total}. Calculate the SS/MS_{total} for each factor and the interaction:

```
> ms_tot=340/15
> sex=256/ms_tot
> day=25/ms_tot
> int=4/ms_tot
```

You might want to confirm that they hold suitable values:

```
> sex
[1] 11.29412
> day
[1] 1.102941
> int
[1] 0.1764706
```

4 Use the function 'pchisq()', or actually one minus this value, to give you the *P*-value associated with these values just calculated and the relevant degrees of freedom (all are one in the example, but will always be $n-1$, one less than the number of factor levels):

```
> 1-pchisq(sex,1)
[1] 0.0007775304
> 1-pchisq(day,1)
[1] 0.2936215
> 1-pchisq(int,1)
[1] 0.6744241
```

5 The results can be laid out as in the SPSS example above. In this case the effect of 'sex' is highly significant as $P=0.0007$, but the other factor and the interaction were not significant. In the parametric version of the test both factors were significant as was the interaction, which demonstrates how weak this test is.

MINITAB There is no way of carrying out this test directly in MINITAB but with a little extra work the test is achievable.

1 Put the observations into a single column then the factors in the next two columns with the factor levels coded as integers. Label the three columns appropriately. From the 'Data' menu select 'Rank…'. Move the data observation column into the 'Rank data in:' box and in the 'Store ranks in:' type a new name (e.g. 'rank'). Click 'OK' and the rank position of the data appears in the new column with the smallest value ranked as 1.

2 Carry out a conventional two-way ANOVA using the 'rank' rather than the original observations. From the 'Stat' menu select 'ANOVA' then 'Two-Way…'. Put the rank column into the 'Response:' box and the two factors into the 'Row factor:' and 'Column factor:' boxes. Click 'OK'

(Or, assuming the ranks are in c4 and the factors in c2 and c3, type 'twow c4 c2 c3' at the MTB> prompt in the session window. Or you can input commands using 'Edit' menu then 'Command Line Editor'.)

The example will give the following output:

Two-way ANOVA: rank versus Sex, Day length

Source	DF	SS	MS	F	P
Sex	1	256	256.000	55.05	0.000
Day length	1	25	25.000	5.45	0.038
Interaction	1	4	4.000	0.87	0.369
Error	12	55	4.583		
Total	15	340			

S = 2.141 R-Sq = 83.82% R-Sq(adj) = 79.78%

---

3 Ignore the last two columns with F- and P-values. Record the sum of squares (SS) values for the two factors and the interaction and calculate the total mean squares (MS; SS_{total}/DF). Then calculate values for SS/MS_{total} either in MINTAB using 'Calc', then 'Calculator...', or by using a calculator. Type the calculated values into a single column in the spreadsheet (make sure to remember which value applies to each factor). Then use MINITAB to look up the values on a chi-square table: from the 'Calc' menu select 'Probability distributions' then 'Chi-square...'. Select the 'Cumulative probability' option. Then input the appropriate degrees of freedom (if the factors and interaction have different degrees of free-dom then you will have to repeat this process for each different number of degrees of freedom) in the example there is one degree of freedom for all three values. Then in the 'Input column:' box put the name of the variable containing the calculated values and choose an empty column for the output 'Optional storage:'. Click 'OK'.

4 So far this has generated values for the probability of getting the number calculated *or lower* whereas the number *or higher* is required for a P-value. So from the 'Calc' menu select 'Calculator...'. In the dialogue window put the name 'pvalue' in the 'Store result in variable:' box. Finally in the 'Expression:' box type '1-' (i.e. one minus) then the name of the column where the chi-square output was sent.

Using the example, this table shows the numbers used:

	SS	MS	SS/MS$_{total}$	d.f.	Cumulative chi-square value	P-value
Day length (factor)	25		1.10	1	0.705734	0.29427
Sex (factor)	256		11.29	1	0.999221	0.00078
Interaction	4		0.176	1	0.325166	0.67483
Total	340	22.67		15		

In the example the P-values for 'day length' and 'interaction' are both well above 0.05 so the null hypothesis that there is no difference between factor levels is accepted. However, the P-value for 'sex' is less than 0.001, indicating that the null hypothesis should be rejected and the two sexes have highly sig-nificantly different food intake.

Excel There is no direct way of carrying out this test in Excel although, providing the design is balanced (there is the same number of observations in each factor combination), it can be done using two-way ANOVA with only a few extra steps to rank the data and look up the significance of the test result.

1 First the raw data should be turned into ranks starting with the lowest value as 1 and then put into a table with row and column headings. So if the raw data were input as before, the observations would be in cells B2–C9. To turn this into ranked information the 'RANK' function is used. In cell E2 type '=RANK(B2,$B$2:$C$9,1)'. This gives the rank position of the datum in cell B2 in the set of 16 numbers. The default in Excel is to rank the largest value as 1. To rank the smallest value as 1, the option '1' is used in the 'RANK' function, as shown. Copy cell E2 to F2 and add labels. The example data set now becomes:

	Long	Short
Female	11	14
	9	12
	10	15
	13	16
Male	1	5
	4	8
	7	2
	3	6

2 Carry out a 'Two way ANOVA with replication' exactly as described above. From the 'Data' menu/ribbon select 'Data analysis…' and then 'ANOVA: Two-Factor with Replication'. Define the range to cover the cells with the ranked data and the group labels. Indicate how many observations there are for each factor combination (four in the example), leave 'Alpha' as 0.05 and send the output to a clear area of the spreadsheet. Click 'OK'.

3 Ignore the output except for the ANOVA table at the bottom. Cut out the first four columns of the output and paste it into the top left corner of a new sheet. For the example the required output is:

Source of Variation	SS	df	MS
Sample	256	1	256
Columns	25	1	25
Interaction	4	1	4
Within	55	12	4.583333
Total	340	15	

Moving the output like this is not required by the package but it does remove distractions.

4 Calculate the total mean square, MS_{total}. This is the total sum of squares (SS) divided by the total degrees of freedom (df). If you have moved the table so that cell 'A1' contains the text 'Source of Variation' then just type into cell D7 '=b7/c7'. The statistic for the Scheirer–Ray–Hare test can now be calculated for the two main factors and the interaction as it is SS/MS_{total}. In column 'e' type '=b2/d7' in row 2, '=b3/d7' in row 3 and '=b4/d7' in row 4. This gives three numbers that need to be looked up on a chi-square table: very easily done in Excel.

5 In column 'f' type 'P-value' in row 1 as a label. Then in row 2 type '=CHIDIST(e2,c2)' and the P-value for the value in column 'e' and the degrees of freedom in column 'c' is given. Copy this into the two cells below (easily done in Excel by selecting cell 'f2' then clicking on the small black square in the bottom right corner of the cell and dragging it down to cover the next two cells).

The final table for the example data is as follows:

	F2			f_x	=CHIDIST(E2,C2)	
	A	B	C	D	E	F
1	*Source of Variation*	*SS*	*df*	*MS*	*SS/MStot*	*P-value*
2	Sample	256	1	256	*11.29412*	0.000778
3	Columns	25	1	25	*1.102941*	0.293622
4	Interaction	4	1	4	*0.176471*	0.674424
5	Within	55	12	4.583333		
6						
7	Total	340	15	22.66667		

The *P*-values for 'Columns' (='day length' in the example) and 'Interaction' are well above 0.05 so the null hypothesis that the factor levels have the same median is accepted. The *P*-value for 'Sample' (='sex' in the example) is well below 0.05. The null hypothesis is rejected and the alternative hypothesis that the two groups (sexes) have different food intakes is accepted. Which level is the higher must be determined by inspection.

There are more than two independent ways to classify the data

If each observation can be assigned to groups using more than two different factors (ways of classifying the data), each of the factors can be divided into groups and assigned integers and, most importantly, the factors are all fully independent of each other, then the data are suitable for multifactorial testing. The biggest danger is thinking that factors are independent when they are not. For example, say you sample from a variety of woodlands and use wood number as a factor then divide the samples into two groups based on whether they were in a northern or southern area. The woodland number is not independent of the region (e.g. wood number 6 can only be in one or the other). If this is the case then a nested design is required rather than a fully factorial one (see below).

Multifactorial testing

If there are more than two factors and they are fully independent of each other then the design is said to be factorial. When there are just two factors there is scope for only one interaction term: factor A with factor B. As more factors are added to the analysis the number of interaction terms that are possible grows rapidly. For three factors there are four interaction terms ($A \times B$, $A \times C$, $B \times C$ and $A \times B \times C$), for four factors there are 11.

Three-way ANOVA (without replication)

If time, space or resources are very limited then this design is quite common. If there are three independent ways of classifying the data into groups and there is only one observation for each factor combination then it is still possible to carry out an ANOVA. Follow the instructions for three-way ANOVA with replication but be aware that there is no way of calculating the interaction between the three factors (this is used as the 'error' by the calculation). This means that only two-way interactions should be allowed in the model.

SPSS Follow the instructions for three-way ANOVA with replication (see next section), but if you have no replication (i.e. there is only one observation for each factor combination) when you have arranged the factors and the dependent variables you must first go to the 'Model...' submenu within the 'Univariate' dialogue. Click on 'Custom'. Then highlight the three factor names and under 'Build terms' select 'Main effects' from the menu before moving the factors across to the 'Model:' box. Then repeat with 'All 2-way' selected. In the example this will give six lines in the 'Model:' box, each effect and three pairwise interactions. Click 'Continue' to return to the 'Univariate' dialogue and 'OK' to run.

R Follow the instructions for three-way ANOVA with replication but remember to specify that the three-way interaction is not included in the model. This means that the three-way ANOVA using 'aov()' or 'lm()' would need to have the model modified by adding a term of '-A:B:C' where A, B and C are factors and the colon indicates interaction. Here is an example of a three-way ANOVA using 'aov()' and then with the three-way interaction removed to allow an unreplicated design:

```
> summary(aov(intake~sex*day*region))
> summary(aov(intake~sex*day*region-sex:day:region))
```

MINITAB Follow the instructions for three-way ANOVA with replication but when specifying the 'Model:' the three-way interaction (e.g. c2*c3*c4) should be omitted. If column names are not used and the grouping variables are in columns 2, 3 and 4 this would give a model statement of 'c2 c3 c4 c2*c3 c2*c4 c3*c4'.

Excel Only one-way and balanced two-way ANOVAs are possible in Excel.

Three-way ANOVA (with replication)

If there are three fully independent ways of dividing the data into groups (e.g. site of collection, species and sex) and there is more than one observation for each factor combination then this design is appropriate. As with all ANOVA the test assumes that the data are continuous and approximately normally distributed and that the variance is approximately equal for each factor level. Unfortunately there is no non-parametric equivalent to use if the assumptions are not met.

An example

A group of agricultural ecologists are interested in the effect of rabbit grazing on plant communities near to warrens. The observations are carried out on two sites on different soils. At each site there are two areas for study, one is 10 m from an active warren and one is 100 m from a warren. In each study area six quadrats are positioned at random. Two quadrats are left untouched as a control, two have caged exclosures placed around them that totally exclude rabbits and two are 'procedural controls' where the digging for constructing an exclosure is carried out but the rabbits are not excluded with fences. A procedural control is required in this experiment as otherwise any effects could be attributed either to the disturbance from construction or the exclusion of rabbits. Procedural controls are often overlooked when designing experiments (see Chapter 4 for a description of procedural control).

The data collected are the standing crop of grass collected 100 days after the exclosures were placed. There are three independent factors: site (two levels, coded 1 and 2), distance from warren (two distances, 10 and 100 m, coded 1 and 2) and exclosure type (three levels, coded 0 for control, 1 for procedural control and 2 for exclosure). There are two observations for each factor combination, giving a total of 24. Here are the data in table format:

		Distance from warren					
		1			2		
		Exclosure type					
		0	1	2	0	1	2
Site	1	112	115	187	141	121	189
		116	102	175	101	157	186
	2	121	145	198	135	141	208
		138	124	168	129	133	206

Output from statistical packages often has more information than is required. The ANOVA table included in a report or publication could be presented as follows:

Source of variation	d.f.	SS	MS	F	P
Distance	1	1218.4	1218.4	6.44	0.026 *
Exclosure	2	6841.4	3420.7	18.07	<0.001 ***
Site	1	590.0	590.0	3.12	0.103
Distance × exclosure	2	327.0	163.5	0.864	0.446
Distance × site	1	1.0	1	0.006	0.942
Exclosure × site	2	58.3	29.2	0.154	0.859
Distance × exclosure × site	2	282.3	141.2	0.746	0.495
Error	12	2271.5	189.3		
Total	23				

This sort of table should always be accompanied with a table caption that explains the test used in more detail, so that the reader does not need to refer to the main text. The asterisks used to indicate levels of significance should also be explained: $*=0.01 < P < 0.05$ etc.

This example assumes that the factors are all fixed. That is, the factor levels are meaningful rather than arbitrary labels (see page 193 for a discussion of fixed and random effects). In this example there is no doubt that 'Distance' and 'Exclosure' are fixed effects as the groups have clear meanings. The factor 'Site' is one that might be either fixed or random. If treated as fixed then the analysis is considering differences between two specified sites. If treated as random then the difference considered is that between any two sites and their location is unimportant.

SPSS The test is essentially the same as for two-way ANOVA. Arrange the data so that there is one column for the observations and one for each of the three factors with the group labels coded as integers. The group labels can be decoded in the 'Variable View' using the 'Values' column, although I have not done this in the example. It will be assumed that the three factors are all fixed effects.

From the 'Analyze' menu select 'General Linear Model' then 'Univariate…'. Put the observations into the 'Dependent:' box and the three factors into the 'Fixed Factor(s):' box. Click 'OK' to run the test. Using the example you get the following output:

Univariate Analysis of Variance

Between-Subjects Factors

		N
Site	1	12
	2	12
Exclosure	0	8
	1	8
	2	8
Distance	1	12
	2	12

Tests of Between-Subjects Effects

Dependent Variable: Grass

Source	Type III Sum of Squares	df	Mean Square	F	Sig.
Corrected Model	23565.333[a]	11	2142.303	10.349	.000
Intercept	524512.667	1	524512.667	2533.878	.000
Site	864.000	1	864.000	4.174	.064
Exclosure	21085.083	2	10542.542	50.930	.000
Distance	888.167	1	888.167	4.291	.061
Site* Exclosure	6.250	2	3.125	.015	.985
Site* Distance	37.500	1	37.500	.181	.678
Exclosure* Distance	166.583	2	83.292	.402	.677
Site* Exclosure* Distance	517.750	2	258.875	1.251	.321
Error	2484.000	12	207.000		
Total	550562.000	24			
Corrected Total	26049.333	23			

a. R Squared = .905 (Adjusted R Squared = .817)

The output confirms the test and then gives the number of observations in each level of the three main effects. It is worth checking this to make sure that the data have been input correctly. Note that each datum will be counted three times in this table.

Next comes the ANOVA table. First the name of the dependent variable is confirmed. Then comes an ANOVA table with some extra lines. The most important lines are for the factor variables, called 'Main Effects' in SPSS ('Distance', 'Exclosure' and 'Site' in the example) and their interactions (three possible two-way interactions and one three-way interaction). The 'df' column gives the degrees of freedom which is one less than the number of factor levels for main effects and calculated by multiplication for interactions (e.g. for 'Distance * Exclosure' there are 2×1=2 d.f.). The sum of squares and mean square are given; mean square is sum of squares/degrees of freedom. The last two columns give the important information. The 'F' column is the F-ratio (the factor mean square divided by the error mean square). The P-value, labelled 'Sig.', is in the last column. In the example the *P*-values for both 'Site' and 'Distance' are slightly above the critical 0.05 level so we accept the null hypotheses that there is no difference between sites or between distances. However, the *P*-value for 'Exclosure' is given as 0.000, which should be reported as $P<0.001$, much less than 0.05, so we reject the null hypotheses and accept the alternative hypotheses that exclosures affect the grazing of rabbits. None of the interaction terms has a *P*-value even close to 0.05.

In this output it is best to ignore the lines labelled 'Corrected Model', 'Corrected Total' and 'Intercept'.

Post hoc *testing in SPSS* The ANOVA has indicated that one of the main effects is significant. If there were only two factor levels, we could easily determine which group was higher by inspecting the means of the data. However, as in the example there are three factor levels for the significant effect a *post hoc* test is required. Repeat the method as above but in the 'Univariate' box click on 'Post hoc'. This brings up a dialogue box with a bewildering array of options. First move the required factor from the 'Factor(s)' box to the 'Post Hoc Tests for:' box. This will activate all the options for *post hoc* tests. All will tend to give the same results, but some will be more conservative than others. I suggest the 'Sidak' method, 'SNK' or 'LSD' (least conservative) test, although you should only conduct an LSD test if the factor is significant.

This is the output from three *post hoc* tests using the example data:

Post Hoc Tests

Exclosure

Multiple Comparisons

Dependent Variable: Grass

	(I) Exclosure	(J) Exclosure	Mean Difference (I-J)	Std. Error	Sig.	95% Confidence Interval Lower Bound	95% Confidence Interval Upper Bound
LSD	0	1	−5.6250	7.19375	.449	−21.2988	10.0488
		2	−65.5000*	7.19375	.000	−81.1738	−49.8262
	1	0	5.6250	7.19375	.449	−10.0488	21.2988
		2	−59.8750*	7.19375	.000	−75.5488	−44.2012
	2	0	65.5000*	7.19375	.000	49.8262	81.1738
		1	59.8750*	7.19375	.000	44.2012	75.5488
Sidak	0	1	−5.6250	7.19375	.833	−25.5538	14.3038
		2	−65.5000*	7.19375	.000	−85.4288	−45.5712
	1	0	5.6250	7.19375	.833	−14.3038	25.5538
		2	−59.8750*	7.19375	.000	−79.8038	−39.9462
	2	0	65.5000*	7.19375	.000	45.5712	85.4288
		1	59.8750*	7.19375	.000	39.9462	79.8038

Based on observed means.
The error term is Mean Square (Error) = 207.000.
*. The mean difference is significant at the 0.05 level.

Immediately followed by

Homogeneous Subsets

Grass

	Exclosure	N	Subset 1	Subset 2
Student-Newman-Keuls[a,b]	0	8	124.1250	
	1	8	129.7500	
	2	8		189.6250
	Sig.		.449	1.000

Means for groups in homogeneous subsets are displayed.
Based on observed means.
The error term is Mean Square(Error) = 207.000.
a. Uses Harmonic Mean Sample Size = 8.000.
b. Alpha = 0.05.

There are two tables. This first gives the LSD and Sidak method results. There are six lines for each method giving each pairwise comparison. This means that each pair appears on the list twice (e.g. as 0 / 1 and 1 / 0). The table gives the difference between the mean observations from the two factor levels, a measure of the standard error of the differences and then, most importantly, a 'Sig.' column. In the example, for both LSD and Sidak methods there is no significant difference between groups 0 and 1, but there is a highly significant difference between groups 0 and 2 and between 1 and 2. This indicates that areas surrounded by exclosure fences have, perhaps unsurprisingly, a much higher mean grass weight.

The second table gives the same information in a different format as the result of the Student–Newman–Keuls (SNK) test. This *post hoc* test puts groups 0 and 1 in a 'homogeneous subset' that is different from group 2. This table also, usefully, gives the mean values of the observations in each group.

R The data should be in a single variable and there are three other variables giving the factor levels of the predictor variables. R will probably need to be told that the factors where levels are coded as numbers should be treated as factors rather than covariates in the model. This is done with the function 'as.factor()'. There are several ways to reach three-way ANOVA, but the functions 'aov()' and 'lm()' are the simplest to use. Assuming the response variable is 'grass' and the predictor variables (factors) are 'exclosure', 'distance' and 'site' the three-way ANOVA with all interactions is:

```
> model<-(aov(grass~exclosure*distance*site))
```

The output of the 'aov()' function is now in 'model' and the output can be seen using 'summary()':

```
> summary(model)
                   Df Sum Sq  Mean Sq F value Pr(>F)
exclosure           2 21085.1 10542.5 50.9302 1.3e-06 ***
distance            1 888.2    888.2  4.2907  0.06054
site                1 864.0    864.0  4.1739  0.06365
exclosure:distance
                    2 166.6     83.3  0.4024  0.67742
exclosure:site      2 6.2        3.1  0.0151  0.98504
distance:site       1 37.5      37.5  0.1812  0.67791
exclosure:distance:site
                    2 517.8    258.9  1.2506  0.32112
Residuals          12 2484.0   207.0
---
Signif. codes:  0 '***' 0.001 '**' 0.01 '*' 0.05 '.' 0.1
 ' ' 1
```

The results give for each factor and interaction (interactions are denoted by ':') the usual degrees of freedom, sum of squares, mean square, F-ratio and P-value (labelled 'Pr(>F)'). Significant lines are indicated with asterisks and a key is provided. Here one of the main effects is highly significant, $P<0.001$, with three asterisks. The other main effects both have P-values close to 0.05, so may warrant further investigation. None of the interactions have P-values anywhere near 0.05.

As the significant factor has three levels we need a *post hoc* test to reveal which factor levels are different from which. The function 'TukeyHSD()' can be used on the same results of the 'aov()' from earlier and the factors to be analysed are put in quotes:

```
> TukeyHSD(model, "exclosure")
Tukey multiple comparisons of means
95% family-wise confidence level

Fit: aov(formula=grass ~ exclosure * distance * site)

$exclosure
      diff     lwr        upr       p adj
1-0   5.625   -13.56694   24.81694  0.7207163
2-0   65.500   46.30806   84.69194  0.0000027
2-1   59.875   40.68306   79.06694  0.0000069
```

The results show that factor level 2 is significantly different from the other two, but that levels 1 and 0 are not significantly different from each other.

MINITAB Put the observations into a single column. Use three further columns for the group labels of the three factors (coded as integers). Label the columns appropriately.

From the 'Stat' menu select 'ANOVA' then 'General Linear Model...'. Move the observation column into the 'Responses:' box and the three factor columns into the 'Model:' box. Unfortunately if interactions are to be investigated these need to be added separately. So, assuming the factors are in columns 2, 3 and 4 then the following needs to be added to the model box: 'c2*c3 c2*c4 c3*c4 c2*c3*c4'. The names of the factors could be used instead of column codes. (If there is no replication – i.e. only one observation for each factor combination – then the final c2*c3*c4 term should be omitted.)

The following output appears:

General Linear Model: grass versus distance, exclosure, site

```
Factor      Type    Levels   Values
distance    fixed   2        1, 2
exclosure   fixed   3        0, 1, 2
site        fixed   2        1, 2

Analysis of Variance for grass, using Adjusted SS for Tests
```

Source	DF	Seq SS	Adj SS	Adj MS	F	P
distance	1	888.2	888.2	888.2	4.29	0.061
exclosure	2	21085.1	21085.1	10542.5	50.93	0.000
site	1	864.0	864.0	864.0	4.17	0.064
distance*exclosure	2	166.6	166.6	83.3	0.40	0.677
distance*site	1	37.5	37.5	37.5	0.18	0.678
exclosure*site	2	6.2	6.2	3.1	0.02	0.985
distance*exclosure*site	2	517.7	517.7	258.9	1.25	0.321
Error	12	2484.0	2484.0	207.0		
Total	23	26049.3				

```
S = 14.3875 R-Sq = 90.46% R-Sq(adj) = 81.72%
```

This is essentially the same as the output for two-way ANOVA only with several more lines. I always advise checking that the first part of the output confirming the factor names, number of groups and the integer codes. The ANOVA table contains lines for each of the factors, then the interactions and finally the error and total. The P-values (labelled 'P') are less then 0.05 for only one of the factors and none of the interactions. The null hypotheses that grass weight comes from the same distribution for all levels of these two factors is rejected and an alternative hypothesis is accepted. However, in this case the two non-significant factors both have P-values very near to 0.05 and further analysis should be considered. If a *post hoc* test is required, as it would be if there are three or more factor levels and the factor has a P-value less than 0.05, go to the 'Comparisons' button before running the ANOVA test.

Visualization of which factor level is higher than which is simple to achieve in MINITAB. From the 'Stat' menu select 'ANOVA' then 'Interactions plot...'. Put the factor columns into the 'Factors:' box (no need for the interaction terms

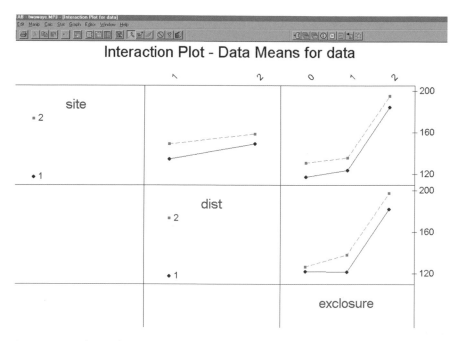

Fig. 7.15 Visualizing three-way ANOVA in MINITAB using the 'Interactions plot'. The factor 'exclosure' has clearly higher means for level 2 when divided by 'dist' or 'site'.

this time) and the raw data in the 'Raw response data in:' box. Click 'OK' and a set of graphs showing the mean values of the raw data for every factor combination appears (see Fig. 7.15). It is clear that the data from level 2 of the exclosure factor is much higher than the other two levels of this factor (see the interaction section above for an explanation of how these graphs can be interpreted).

(Or, if the command interface is enabled, and assuming observations in c1, factors in c2, c3 and c4 'GLM c1=c2 c3 c4 c2*c3 c2*c4 c3*c4 c2*c3*c4' at the MTB> prompt in the session window.)

Excel There is no direct way to carry out this test in Excel.

Multiway ANOVA

If there are more than three ways of dividing the data into groups and each of the classifications is independent of the others then ANOVA may be carried out. These multifactorial designs become increasingly difficult to interpret as there is an explosive increase in the number of interaction terms as the number of factors goes up. Furthermore, unless an experiment is designed to be fully factorial there are likely to be combinations of factors where there are no observations.

Finally, once there are many factors the chance that they are all independent of each other diminishes. Each factor should be considered in turn: is it independent of all other factors? Can two factors be sensibly combined in some way? Is one factor nested within another (see next section)? Is one factor more appropriately investigated as a covariate (see section on analysis of covariance, page 238)?

Not all classifications are independent

A very large number of experiments that are treated as if all the factors are independent prove to be, on further investigation, nothing of the sort. Therefore it is important that the independence of the factors in the analysis is considered carefully.

Non-independent factors

Some ways of classifying data are always going to be fixed and therefore main factors in an analysis. For example, a commonly used factor in biology is 'sex': there are two sexes (although in some species there are intermediate 'inter-sexes' or hermaphrodites) and once an individual is labelled as either male or female that classification is fixed. Other factors are not so clearly fixed: if a set of observations is divided into two equal-sized groups by weight then an individual may find itself in the heavy group at the start but later moved into the light group. This type of factor is not really fixed although the label 'heavy' really means something about the observation. Consider a study on blood pressure in humans using the two factors I have described here: 'sex' and 'weight'. Human males are heavier than females and therefore there will be more males in the heavy group therefore *the two factors are not independent of each other*. This type of independence can be checked using a chi-square test of association considered in the next chapter (page 199). If the factors are not independent then two-way ANOVA is inappropriate. One solution is to apply an analysis of covariance (ANCOVA) using 'sex' as a fixed factor and 'weight' as a covariate. This analysis is considered in the next chapter (page 238).

Nested factors

Nested factors are never 'fixed', meaningful factors, they are always factors numbered for convenience where the number given does not really mean anything. For example, in a greenhouse experiment there may be plants in numbered pots with four pots per numbered tray and three trays per bench with two benches in each of two greenhouses. At each level 'pot', 'tray', 'bench' and 'greenhouse' the numbers are only used for convenience. These are called random factors. In this example, the factor 'pot' is said to be nested within 'tray', the 'trays' nested within 'bench' and 'bench' within 'greenhouse'. It is important that the numbering system reflects this nesting: pots should be labelled 1–4 in *each* tray and trays labelled 1–3 in each greenhouse. *Nested factors are always random but not all random factors need to be nested* (e.g. the factor 'site' in the three-way ANOVA example given above might be a random factor, but it is not nested). *Post hoc* tests may not be carried out on nested factors.

Random or fixed factors

I have touched on the difference between 'fixed' or meaningful factors and 'random' or convenience factors above. There is, however, not a strict distinction between the two types of factor and the same factor can be treated as 'fixed' or 'random' in different tests. For example, in an analysis of behaviour in *Drosophila melanogaster* mutants the mutants used could be either a fixed or random factor. If the mutant is treated as fixed (so that *vestigial* is coded as 1 and *white-eye* as 2) then this implies that the significance level of any difference is due to the characteristics of the particular named mutants. However, if the factor is random then any significance implies differences between *any* two randomly selected mutants.

The next section considers designs with a simple hierarchy of nested factors. More complicated arrangements where it is appropriate to test both nested and factorial designs combined are surprisingly common. Such designs are extremely widely used in biology but they are often very difficult to achieve in the statistical packages, although it must be said that in R crossed and nested designs are equally easy to implement.

Nested or hierarchical designs

If the experiment or sampling design has only one 'random' factor at the top of the hierarchy and all other factors are 'nested' within it then the design is said to be a pure nested one. If the top level is 'fixed' and there are other factors nested within it the design is often called a 'mixed model'. There is no limit to the number of factors that can be nested in this way. The simplest case of a single fixed factor and one nested is considered here.

Two-level nested-design ANOVA

An analysis with two factors. The standard factor in this type of design may be 'fixed' or 'random'. There should then be a second, 'random' factor nested within the standard factor.

An example

In an experiment on cholesterol levels in mouse blood two levels of fat intake are fixed by the researchers and coded as levels 0 and 1. For each level of 'intake' there are three populations of mice in separate cages and from each of these cages three individual mice are selected at random for blood testing. The six populations of mice are assigned to the factor 'cage', which is a random factor nested within 'intake' with the three cages in each group coded 1, 2 and 3 (clearly these labels have no real meaning and are only used for convenience). There are six cages in all but they must not be coded as 1 to 6.

The data collected are as follows:

		Intake				
		0			1	
			Cage			
Mouse	1	2	3	1	2	3
1	55.6	62.5	58.9	85.6	68.5	65.6
2	62.5	68.3	54.2	98.2	69.3	71.0
3	68.2	58.2	63.5	75.1	88.2	78.3

Note: a common error made when analysing this type of data is that 'mouse' is treated as a factor. This might seem sensible as the mice are labelled 1–3. However, it is inappropriate as there is only one observation per mouse, ANOVA would have no variation to work with if it was used as a factor. Therefore 'mouse' is actually the level of replication within the factor 'cage'. If there were two observations per mouse then mouse could be a further random factor that would be nested within 'cage'.

Using the example data the following ANOVA table should be used in a report:

Source of variation	df	Sum of squares	Mean square	F-ratio	P-value
Intake	1	1215.24	1215.24	18.90	0.001 **
Cage within intake	4	377.42	94.35	1.47	0.272
Error	12	771.68	64.31		
Total	17				

SPSS Nested analyses are achievable, but slightly awkward, in SPSS. Put the observations in a single column. Use further columns for the factors with the levels coded as integers. From the 'Analyze' menu select 'General Linear Model' then 'Univariate…'. Put the observations into the 'Dependent variable:' box and the fixed factor in the 'Fixed Factor(s):' box ('intake' in the example is a fixed effect). Put the nested factor in the 'Random Factor(s):' box ('cage' in the example). (A detour via the 'Options…' button or the 'Plots…' button can be useful as means by factor levels can be requested allowing an easy comparison of groups).

Now the awkward bit…click on 'Paste'. This brings up the command line 'Syntax' window of SPSS. It will probably be called 'Syntax1'. For the example it contains the following:

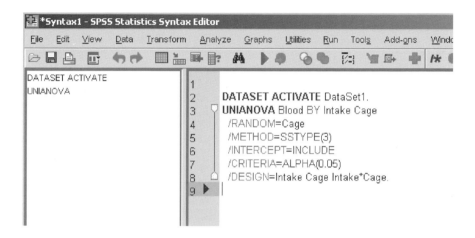

To carry out a nested ANOVA the '/DESIGN' line must be altered to specifically request it. The nested factor must be specified as being within the main factor. The nested factor is listed before the factor it is nested within and, rather counterintuitively, the fixed factor has to be in parentheses. In the example the last line becomes '/DESIGN cage(intake) intake.'. To run this design from the 'Syntax' window click on the 'play' button (an icon of a small blue triangle pointing to the right).

The following output appears:

Univariate Analysis of Variance

Between-Subjects Factors

		N
Intake	0	9
	1	9
Cage	1	6
	2	6
	3	6

Tests of Between-Subjects Effects

Dependent Variable: Blood

Source		Type III Sum of Squares	df	Mean Square	F	Sig.
Intercept	Hypothesis	87041.827	1	87041.827	922.498	.000
	Error	377.418	4	94.354[a]		
Intake	Hypothesis	1215.245	1	1215.245	12.880	.023
	Error	377.418	4	94.354[a]		
Cage(Intake)	Hypothesis	377.418	4	94.354	1.467	.272
	Error	771.680	12	64.307[b]		

a. MS(Cage(Intake))
b. MS(Error)

First comes a table confirming the number of observations for each factor level. This is worth checking to confirm it matches your expectations. It can often highlight an error in data entry.

The ANOVA table confirms the name of the dependent variable. Then comes a conventional ANOVA table with slightly more information than is needed. The important lines are those labelled with the names of the main effect and nested factor and the bottom error (or residual) line. In the example the P-values (labelled 'Sig.') show that there is no effect of 'cage(intake)' (i.e. 'cage' nested within 'intake') but there is a significant effect of 'intake' as P is less than 0.05. The degrees of freedom for nested factors are slightly odd as they have one less than the number of levels for *each* level of the main factor (in the example this is number of cages: 1 for each intake or $(3-1) \times 2 = 4$).

The final table gives the expected mean squares. These are very important when calculating ANOVA by hand, but are not useful here.

A possible complication: note that in this case SPSS has generated a different F-ratio to that shown in the table above. This is because the calculation of F-ratios in nested ANOVA is something of an art. In this case SPSS has used the mean square of the nested factor as the denominator for the F-ratio while the table above, R and Minitab have used the mean square of the error instead. There is a convention that the mean square of the nested factor is used as the denominator *whenever the factor is significant* but in other circumstances there is debate over the best denominator to use. So, if the nested factor has a P-value less than 0.05 it is customary to use the mean square of this factor as the denominator to calculate the F-ratio of the main factor rather than the error mean square. SPSS 10 appears to conform to this convention, but more recent versions of SPSS do not. It is worth checking the F-ratios carefully if you have a different version of the package. With more complicated designs selection of the correct denominators can become a rather long-winded process.

R Nested ANOVA designs are very easy to specify in the function 'aov()' in R. Just use the operator '/' to indicate that a factor is nested in another: 'aov(data~A/B)' indicates that factor 'B' is nested in factor 'A'. The data need to be in a single variable and each of the coding variables in separate and clearly labelled variables. It is important to realize that although there are six cages in this design there will only be three different labels as there are three cages in each level of 'intake'. Using the example data with observations in 'cholest' and 'cage' and 'intake' as the factors and ANOVA with 'cage' nested within 'intake' is:

```
> summary(aov(cholest~intake/cage))
              Df  Sum Sq   Mean Sq  F value  Pr(>F)
intake        1   1215.24  1215.24  18.8977  0.0009496 ***
intake:cage   4   377.42   94.35    1.4673   0.2724820
Residuals     12  771.68   64.31

---
Signif. codes: 0 `***' 0.001 `**' 0.01 `*' 0.05 `.' 0.1
` ' 1
```

R indicates in the output that a factor is nested by 'A:B' indicating that 'B' is nested within 'A'. In this case 'cage' is nested within 'intake'. The output comprises a normal ANOVA table with degrees of freedom, sum of squares, mean square, *F*-ratio and *P*-value given. Asterisks highlight significant lines. Here the effect of 'intake' is highly significant, $P<0.001$.

MINITAB Put the observations in a single column. Use additional columns for the factors with the levels coded as integers or text labels. From the 'Stat' menu, select 'ANOVA' then 'General Linear Model...'. Move the observation column into the 'Responses:' box. Then put the main factor and the nested factor ('intake' and 'cage' in the example) into the 'Model:' box. To show that the nested factor is nested it must be followed by the name of the main factor in parentheses (in the example the 'Model:' box reads: 'intake cage(intake)'). Click 'OK'. (A detour via the 'Options...' button can be useful as means by factor levels can be requested allowing comparison of groups). You might note that there is a 'Fully Nested ANOVA' option within the ANOVA model, but I find that this is much less controllable than using 'General Linear Model' in MINITAB. If you use the 'Fully Nested' approach you should put the main effect into the 'Factors:' box first then the nested factor. There is no need for the syntax with parentheses.

(Or, if the command interface is enabled, and assuming data in c1, main factor in c2 and nested factor in c3, type 'GLM c1=c2 c3(c2)' at the MTB> prompt in the session window.)

You get the following output from the example.

```
General Linear Model: cholest versus intake, cage

Factor          Type    Levels   Values
cage (intake)   fixed      6     a, b, c, a, b, c
intake          fixed      2     high, low

Analysis of Variance for cholest, using Adjusted SS for Tests

Source          DF    Seq SS   Adj SS   Adj MS      F       P
cage (intake)    4    377.42   377.42    94.35    1.47   0.272
intake           1   1215.25  1215.25  1215.25   18.90   0.001
Error           12    771.68   771.68    64.31
Total           17   2364.34

S = 8.01914 R-Sq = 67.36% R-Sq(adj) = 53.76%
```

The output confirms the test used, the names of the factors, how many levels there are for each factor and the integer labels used (or text labels as in this example). Check these are correct before looking at the ANOVA table. The table itself is very similar to the 'ideal' table given in the example. The degrees of freedom may seem a little odd in a nested analysis. The main factor will have one fewer degrees of freedom than the number of factor levels as usual. The nested factor will have one less than the number of groups *for each level of the main factor*. In the example there are three cages so there are two degrees of freedom for each of two levels of the main factor to give four degrees of freedom. The important column of the table is the final one with the *P*-values (in the example there is no significant effect of 'Cage(Intake)'; that is, cage nested within intake, but a highly significant result for 'Intake').

Excel There is no easy way to carry out this type of nested analysis in Excel.

The tests 2: tests to look at relationships

Is there a correlation or association between two variables?

Observations assigned to categories

If observations are given a qualitative value then the data are said to be categorical (i.e. they have been assigned to categories). There are many occasions when it is not possible to express data in any other way (such as when scoring flower colours or species). There will be other circumstances when the categories are for convenience only, and are achieved by imposing a set of categories on a continuous scale. This can either be an arbitrary scale (e.g. the effect of an illness from 'well' through 'showing symptoms' to 'dead') or a scale that could be measured (e.g. dividing tree heights into 'short', 'medium' and 'tall').

However, if the observations are assigned to categories in this way there are several tests that can be applied to determine whether the division of the observations into classes is independent or not. In other words 'are the observations for two categorical variables associated?'. The **chi-square test of association, phi coefficient** and **Cramér coefficient** are considered here.

Chi-square test of association

This is one of the most widely used statistical tests of all. It is beguilingly simple and has few underlying assumptions. In fact, the test is so simple to carry out it is often not properly supported in statistical packages leaving the user to do a lot of the work. If observations can be assigned to one of two or more categories in two variables then chi-square in appropriate. The null hypothesis is that the categories in the two variables are independent (i.e. the category an observation is assigned to for one variable has no effect on the category it is assigned to for the second variable). For example if 'eye colour' and 'sex' are the two variables and individuals are assigned to either 'blue' or 'brown' and to either 'male' or 'female' then the null hypothesis is that there is no association between sex and eye colour. As there are no assumptions made about the form of the data it is a non-parametric

Choosing and Using Statistics: A Biologist's Guide, 3rd Edition. By Calvin Dytham.
Published 2011 by Blackwell Publishing Ltd.

test although it is rarely described as such (probably because there is no parametric equivalent). (*Note*: it is important to realize that if a continuous variable is forced into a small number of categories then information is lost when the test is calculated and another measure of association is probably more appropriate.)

The chi-square test works by adding up the squared differences between the expected number of observations in a category combination and the actual observed number. The result is then looked up on a chi-square table with a number of degrees of freedom equal to one less than the number of rows multiplied by one less than the number of columns.

Expected values are calculated very simply by putting the data into a table then totalling the observations in each row and column. The expected value for each category combination is the row total multiplied by the column total divided by the total number of observations.

There are complications. If there are expected values lower than one the test should not be used (categories should be combined to avoid this problem). No more than 20% of the expected values should be less than five. An alternative test where it is impractical to raise all the expected values above five is the Fisher's exact test.

Chi-square tests should never be carried out on percentages or data transformed in any way. It must be carried out on frequencies (numbers of observations). When reporting the results in text it is usual to give the statistic, the degrees of freedom and some measure of the *P*-value (e.g. $\chi^2 = 15.62$, d.f. = 4, $P < 0.01$).

An example

A group of students interested in aquatic invertebrates want to determine, in the shortest possible time, whether stream velocity is related to plant growth and substrate. A qualitative survey of 50 randomly assigned stream sections in the study area was carried out. Stream velocity was scored as 'slow' or 'fast' (if more time were available then an accurate measure of the stream velocity could be made, although it might not be very useful as such a measure is very dependent on weather conditions). The stream bed was assigned to one of four categories: 'weed-choked', 'some weeds', 'shingle' and 'silt' (this is one of many possible classification systems that could be used).

The data were collected and compiled into a table as follows; numbers indicate the number of sites with each combination of categories.

		Stream-bed category			
		Choked	Weeds	Shingle	Silt
Velocity	Slow	10	8	2	7
category	Fast	2	6	10	5

The null hypothesis, H_0, is that the proportion of streams in each of the stream bed categories is the same for the two velocity categories. The alternative hypothesis, H_1, is that the proportions vary between the two categories; that is, there is some association with particular stream-bed categories and velocity categories appearing together more often than would be expected by chance and others appearing less often. The test will not determine which category combinations are more or less common than expected although this can be investigated by inspection of the table of raw data and expected values.

The data table needs to be processed to obtain the expected values. First the row and column totals are required.

	Choked	Weeds	Shingle	Silt	Totals
Slow	10	8	2	7	27
Fast	2	6	10	5	23
Totals	12	14	12	12	50

Then the expected value for each of the cells can be calculated. For 'choked' and 'slow' the expected value is the row total (27) multiplied by the column total (12) divided by the total number of observations (50). Of course this sort of manipulation is very easy to do in a spreadsheet such as Excel.

The expected values are:

	Choked	Weeds	Shingle	Silt
Slow	6.48	7.56	6.48	6.48
Fast	5.52	6.44	5.52	5.52

In this example, the value of the chi-square test is 11.036, there are 3 degrees of freedom and the P-value is just over 0.01. This might be reported in the text of a results section as 'a chi-square test showed there was an association between stream velocity category and stream bed category ($\chi^2 = 11.036$, d.f. $= 3$, $P < 0.05$)'.

SPSS The chi-square test of association is not easy to achieve from a table of frequencies. However, if the data are arranged so that each individual is represented by a row with entries in two columns for the two factors then it is very easy to carry out.

Arrange the data in two columns with one column for each of the factors. Input the categories for each observation on a separate row with each categories coded with integers starting at 1. The columns should be labelled appropriately.

The example data will have 50 rows and two columns set out using the following style:

	Velocity	Bed_Cat
1	1	4
2	2	2
3	2	3
:	:	:
50	1	2

The integer codes can be used as labels for the category names. This can be done by clicking on the 'Variable View' and then clicking on the ellipsis ('…') in the 'Values' column. In the 'Value Labels' box that appears labels can be added for each category in turn. After assigning labels for all the categories click 'OK' and repeat for the other variable. Doing this before running the chi-square test will make the output much easier to interpret.

To run the test go to the 'Analyze' menu, choose 'Descriptive statistics' and then 'Crosstabs…'. Move the column containing the 'row' information into the 'Row(s):' box (in the example this was 'velocity') and the other column into the 'Column(s):' box. Then click on the 'Statistics…' button. In the options box that appears select 'Chi-square' and click 'Continue'. Then click on the 'Cells…' button. In the options box select both 'Observed' and 'Expected' in the 'Counts' area. Click 'Continue' then 'OK' in the 'Crosstabs' box. Selecting the 'Display clustered bar charts' produces a useful visual summary of the frequencies in each of the categories, but the default colours are quite awful.

The following output will appear in the 'Output' window:

Crosstabs

Case Processing Summary

	Cases					
	Valid		Missing		Total	
	N	Percent	N	Percent	N	Percent
velocity* bed_cat	50	100.0%	0	.0%	50	100.0%

Velocity* Bed_cat Crosstabulation

			Bed_cat				
			Choked	Weeds	Shingle	Silt	Total
velocity	Slow	Count	10	8	2	7	27
		Expected Count	6.5	7.6	6.5	6.5	27.0
	Fast	Count	2	6	10	5	23
		Expected Count	5.5	6.4	5.5	5.5	23.0
Total		Count	12	14	12	12	50
		Expected Count	12.0	14.0	12.0	12.0	50.0

Chi-Square Tests

	Value	df	Asymp. Sig. (2-sided)
Pearson Chi-Square	11.036[a]	3	.012
Likelihood Ratio	11.945	3	.008
Linear-by-Linear Association	3.160	1	.075
N of Valid Cases	50		

a. 0 cells (.0%) have expected count less than 5. The minimum expected count is 5.52.

The first table just confirms how many rows of data have been processed by the test. The second table gives the data laid out as a contingency table. If labelling of the values for each variable has not been carried out there will only be numbers for the rows and columns. The table gives the number observed above the number expected for each category combination. Row and column totals are also given. Then comes a small table of the results of the test itself. The important line is that labelled 'Pearson Chi-Square'. The number in the 'Value' column is the value of the X^2 approximation to χ^2 (11.036 in the example). Then the degrees of freedom are given (labelled 'df'), equal to the number of rows minus one multiplied by the number of columns minus one. The P-value (labelled 'Asymp. Sig. (2-sided)') appears in the last column. Ignore the other lines in this table. In the example the P-value is less than 0.05 so the null hypothesis is rejected. Inspection of the table shows that the biggest difference between observed and expected values are in the 'Shingle' column, suggesting that there is an association of shingle stream beds with fast-flowing water.

The minimum expected frequency, given as 5.52 in the example, is an important number to look at because if it is less than one then the chi-square test should not be carried out. Caution should also be applied if more than 20% of the cells have expected frequencies less than five. As shown below, SPSS gives a warning if there is a problem (although it still reports the results of the test and gives a *P*-value).

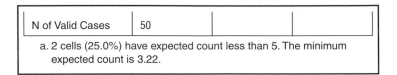

N of Valid Cases	50		
a. 2 cells (25.0%) have expected count less than 5. The minimum expected count is 3.22.			

R The chi-square test is very easy to achieve from a table of frequencies in R. As there will be few numbers to input it is often quickest to input the data at the command line. Here the data from the example are added to dataframe labelled 'stream' and then visualized. It is important to add the data by going down each column in turn (i.e. in the example the second number to be added is 2, not 8):

```
> stream<-matrix(c(10,2,8,6,2,10,7,5),nrow=2)
> stream
      [,1] [,2] [,3] [,4]
[1,] 10    8    2    7
[2,] 2     6    10   5
```

Note how R defines the elements of the matrix with rows labelled first. Element [2,3] would be row 2, column 3. This matrix can then be used in the function 'chisq.test()':

```
> chisq.test(stream)

Pearson's Chi-squared test

data: stream
X-squared=11.0363, df=3, p-value=0.01153
```

The output confirms the test and then the source of the data. The value of chi-square is labelled 'X-squared', the degrees of freedom ('df') and the *P*-value associated with these values. Here $P<0.05$ so the null hypothesis is rejected. Inspection of the table suggests that shingle beds are more frequently associated with fast-flowing water.

If any of the expected values in a chi-square test are less than five R will generate a warning, but will still execute the test:

```
Warning message:
In chisq.test(s) : Chi-squared approximation may be
incorrect
```

If the data are not already in frequency table format that is no problem in R. Imagine the example data are in a list of observations in a text file called 'stream. txt' which looks like this with the variable labels at the top of the columns:

```
Bed          Flow
choked       slow
silt         fast
weeds        slow
...                    ...
```

This can be read into R as 'str' and attached:

```
> str<-read.table("c:\\temp\\stream.txt",header=TRUE)
> attach(str)
```

This can be visualized in a frequency table with the 'table()' function:

```
> table(str)
Flow
Bed          fast       slow
choked       2          10
shingle      10         2
silt         5          7
weeds        6          8
```

Or a chi-square test can be done directly on the list of observations by combining the 'chisq.test()' function and the 'table()' function:

```
> chisq.test(table(str))

Pearson's Chi-squared test

data: table(str)
X-squared=11.0363, df=3, p-value=0.01153
```

MINITAB The chi-square test is easy to carry out in from a table of counts of observations. First input the table exactly as it is laid out in the example. The columns can be labelled with the names of the categories but there is no way to label the rows. Then from the 'Stat' menu, select 'Tables' then 'Chi-Square Test (Two-Way Table in Worksheet) …'. In the dialogue box highlight the relevant columns and click on 'Select' (or double click on the column names). This moves the columns into the 'columns containing the table:' box. Click 'OK'.

The example data with column labels gives the following output in the session window:

Chi-Square Test: Choked, Weeds, Shingle, Silt

```
Expected counts are printed below observed counts
Chi-Square contributions are printed below expected counts

        Choked    Weeds    Shingle    Silt    Total
   1       10        8         2        7       27
          6.48     7.56      6.48     6.48
          1.912    0.026     3.097    0.042

   2        2        6        10        5       23
          5.52     6.44      5.52     5.52
          2.245    0.030     3.636    0.049

Total      12       14        12       12       50

Chi-Sq = 11.036, DF = 3, P-Value = 0.012
```

The output confirms the test then gives a repeat of the table of frequencies but with the addition of row and column totals as well as the expected values, and gives the value for: (observed–expected)^2/expected for each cell in the table. The value of 'Chi-Sq' gives the total of these values, which is the statistic X^2 (estimating χ^2). The degrees of freedom ('DF') is given and then the P-value, labelled 'P-Value'. In the example the P-value is less than 0.05 so the null hypothesis is rejected. There is some association between stream-bed category and velocity category. Inspection of the 'ChiSq' values shows that the biggest contributors are in column 3 ('shingle') and show that a shingle bed is far more common in a fast-flowing stream (row 2) and far less common in a slow-flowing one (row 1) than expected if there were no association between the factors.

Excel The spreadsheet capabilities of Excel make the chi-square test quite easy to carry out.

1 Input the frequency data (number of observations in each category combination) as a table with row and column labels exactly as in the example.

2 Add labels for an extra row and column of totals. In the relevant cells type '=SUM' and the cell numbers required (in the example the row totals column is F so the command in cell F2 is '=SUM(b2:e2)' and the first of the column totals is in B4 with the command '=SUM(b2:b3)'). Once the first cell has been calculated for row and column it can be pasted to calculate the other totals. An easy way of doing this is to highlight the cell, click and hold the small black square in the bottom right corner then drag across to highlight the cells to paste into. Make sure the total of row totals is calculated as well, as this is the total number of observations (50 in the example).

3 A second table, the same size as the first, should be labelled elsewhere on the same sheet. This is for the expected values. Once the row and column labels have been created the expected values for each cell can be calculated as '=row total * column total/number of observations' inputting the relevant cell numbers as required. In the example the top right cell expected value is calculated as '=b4*f2/f4'.

(It is useful to use the $ as an anchor when calculating the expected values. Any cell code containing a $ will *not* alter when the cell contents are pasted elsewhere on the spreadsheet. So when inputting the first cell calculation use '=b$4*$f2/f4' and then paste the cell across the whole table. Paste the cell across four columns and then, with four cells highlighted, paste down a row. The row containing column totals and the column containing the row totals as well as the cell containing the total number of observations (F4) all remain fixed.]

This should be on the spreadsheet:

E8			f_x	=E$4*$F3/$F$4	
A	B	C	D	E	F
Observed	Choked	Weeds	Shingle	Silt	Totals
Slow	10	8	2	7	27
Fast	2	6	10	5	23
Totals	12	14	12	12	50
Expected	Choked	Weeds	Shingle	Silt	
Slow	6.48	7.56	6.48	6.48	
Fast	5.52	6.44	5.52	5.52	

None of the expected values are below five so the test can proceed. If any cell had a value less than one or more than 20% have values less than five then categories should be combined before the test is carried out.

4 Once the observed and expected tables are complete, select an empty cell anywhere on the spreadsheet where the two tables are still on the screen. Then go to 'Insert function' (f_x button), select 'Statistical' and then 'CHITEST' from the list. Click 'OK'. Then put the cell codes for the observed data (not labels) into the 'Actual_range' box. This can be done either by typing in the cell codes for top left and bottom right directly with a colon (:) between them or by selecting the 'Actual_range' box and then clicking and dragging over the cells. Do the same for the 'Expected_range' box selecting the expected values. In the example the 'Actual_range' was B2:E3 and the 'Expected_range' was B7:E8. Then click 'OK'. The *P*-value for the test appears, here the *P*-value is '0.0115' which is lower than the critical '0.05' value so the null hypothesis is rejected. There is an association of 'stream-bed' category with 'velocity category'. Comparison of the observed and expected values indicate than 'Slow' and 'Choked' are more common than would be expected by chance as are 'Fast' and 'Shingle'.

5 It is usual to present the results of a chi-square test giving the degrees of freedom and the value of χ^2. Degrees of freedom for the test is one less than the number of rows multiplied by one less than the number of columns (in the example (4−1)*(2−1)=3). Excel provides a simple method for determining the χ^2 value from a known *P*-value and degrees of freedom. Select an empty cell. Click on the 'Insert function', select 'Statistical' then 'CHIINV'. Click on 'OK'. Input the *P*-value into the 'probability' box (or click on the cell used in step 4 above) and input the degrees of freedom, then click 'OK'. The χ^2 value appears in the cell (11.03 in the example). When reporting the results the chi-square value, the degrees of freedom and the *P*-value should all be given.

Cramér coefficient of association

The Cramér coefficient of association is a test carried out on tables of frequencies in conjunction with a chi-square test that provides additional information about the strength of the association. The statistic X^2 is used to determine significance while the Cramér coefficient (C) is a measure from 0 (no association) to 1 (perfect association) that is independent of the sample size. If this statistic is available as an option on the package you are using then it will allow direct comparison of the degree of association between tables. The Cramér coefficient is the same as the phi coefficient for 2×2 tables.

SPSS Follow the instructions as for the Chi-square test of association. In the 'Statistics' dialogue box within 'Crosstabs' select 'Phi and Cramér's V' in addition to 'Chi-square'. There will be an extra box in the output called 'Symmetric Measures' which contains both statistics and a *P*-value labelled 'Approx. Sig.'.

R For 2×2 tables this can be quite easily calculated using the matrix notation in R where '[2,1]' refers to element in row 2, column 1. First the data is entered directly as a table of frequencies, this is then visualized, next the elements of the array are placed in variables to save space later and finally the phi coefficient is calculated and output:

```
> x<-matrix(c(10,2,8,6),nrow=2)
> x
      [,1]   [,2]
[1,] 10      8
[2,] 2       6
> a<-x[1,1]
> b<-x[1,2]
> c<-x[2,1]
> d<-x[2,2]
> e<-x[1,1]+x[1,2]
> f<-x[2,1]+x[2,2]
> g<-x[1,1]+x[2,1]
> h<-x[1,2]+x[2,2]
> phi<-((a*d)-(b*c))/(sqrt(e*f*g*h))
> phi
[1] 0.2828895
```

MINITAB There is no direct method for performing the Cramér coefficient of association in MINITAB.

Excel There is no direct method for performing the Cramér coefficient of association in Excel.

Phi coefficient of association

This is a special case of the Cramér coefficient for 2×2 tables (i.e. there are only two categories for each of the two variables). It is used in combination with the chi-square test as it generates a value (r_ϕ) that ranges from 0 to 1, indicating a range from no association through to perfect association. It can be calculated directly after a chi-square test has been carried out as it is equal to the square root of the value of X^2 after it has been divided by the number of observations.

SPSS Follow the instructions as for the chi-square test of association. In the 'Statistics' dialogue box within 'Crosstabs' select 'Phi and Cramér's V' in addition to 'Chi-square'.

R Follow the instructions for 2×2 tables given above.

MINITAB There is no direct method for calculating the Phi coefficient of association in MINITAB, but it is a fairly easy calculation to achieve on a calculator or in Excel. In MINITAB it is relatively simple to perform the calculation. Note down the value of X^2 and the number of observations in total. Enable the command prompt by selecting the 'Session' window, then from the 'Editor' menu select 'Enable commands'. At the MINITAB prompt type 'LET K1=value1', where 'value1' is the X^2 from the test. Then 'LET K2=value2', where 'value2' is the number of observations. To calculate the phi coefficient, type 'LET K3=SQRT(K1/K2)'. To display the result, type 'Print K3'.

Excel If the value of X^2 is in cell E1 and the number of observations is in C3, then the formula '=SQRT(E1/C3)' will give the phi coefficient. Alternatively, if the 2×2 contingency table is laid out in cells A1, A2, B1 and B2 then the phi coefficient can be calculated directly by the rather horrible-looking formula:

```
=((A1*B2)-(A2*B1))/(SQRT((A1+B1)*(A2+B2)*(A1+A2)*(B1+B2)))
```

Observations assigned a value

If each individual observation is assigned a meaningful numerical value in two variables there are several tests that may be applied to determine whether the two sets of observations are associated or correlated, the strength of the correlation and whether it is significant or not. Four tests are considered here: the **Pearson's product-moment correlation**, the **Spearman's rank-order correlation**, the **Kendall rank-order correlation** and **regression**. The most appropriate test is determined by the distribution and quality of the of data.

'Standard' correlation (Pearson's product-moment correlation)

When a correlation is mentioned it is almost invariably the Pearson product-moment correlation that is in mind. The statistic, r, to estimate the true correlation, ρ (rho), produced by the test ranges from -1, through 0, to 1 and describes a range of associations between two variables from perfect negative correlation, through no correlation to perfect positive correlation. This test is very widely applied; perhaps too widely, as it has some rather severe assumptions about the distribution of the two variables being investigated. Both variables must be measured on a continuous scale and both must be normally distributed (often termed a bivariate normal distribution). If these assumptions do not apply the Spearman's rank-order correlation should be used instead.

When quoting the results of this test in a report it is usual to use a scattergraph (without a trend line) and the form 'Pearson product-moment correlation indicates a significant positive association between x and y ($r=0.51$, d.f.$=22$, $P<0.05$)'.

Two words of caution:
1 It is quite rare to find two variables that are normally distributed and therefore suitable for Pearson's correlation. Test the data to see if they follow a normal distribution. Consider the alternatives.
2 The statistical significance of correlation is not a good guide to the *real* significance of the correlation. With large sample sizes the value of r required to achieve statistical significance (i.e. to show that there is some relationship between the two variables) is rather low. It is perhaps better to use the value of r^2 as an indicator of the real significance as this value shows the amount of variation in one variable explained by the other.

An example

A marine biologist working on Adélie penguins (*Pygoscelis adeliae*) has measured the sizes of birds forming pairs. The measure used is the length of a bone in the leg which is known, from previous studies, to be a good indication of size. It is measured to the nearest 0.1 mm. The null hypothesis is that male size is not correlated with female size. Unfortunately data could only be collected from six pairs.

Pair	Female	Male
1	17.1	16.5
2	18.5	17.4
3	19.7	17.3
4	16.2	16.8
5	21.3	19.5
6	19.6	18.3

It is assumed that both variables are normally distributed. This data set is a little small to test this although a larger sample of the population might be more suitable (see 'Do frequency distributions differ?' in the previous chapter, page 72). In this case the null hypothesis is rejected as there is a significant positive correlation between male and female size indicating that there is positive, assortative mating in this species. The value of r is 0.88 and r^2 is 0.77 indicating that 77% of the variation in the size of one sex is explained by the size of the other.

SPSS This is one of the easiest tests to carry out in SPSS. Input the data in two columns and add appropriate column labels. The cases (pairs in the example) do not require a separate column. From the 'Analyze' menu, select 'Correlate' then 'Bivariate…'. In the dialogue box move the names of the two variables into the 'Variables:' box and make sure that the 'Pearson' option and 'Two-tailed' are checked. Click 'OK'.

The following output appears:

Correlations

Correlations

		Female	Male
Female	Pearson Correlation	1	.881*
	Sig. (2-tailed)		.020
	N	6	6
Male	Pearson Correlation	.881*	1
	Sig. (2-tailed)	.020	
	N	6	6

*Correlation is significant at the 0.05 level (2-tailed).

The output gives a matrix of correlation coefficients, degrees of freedom and P-values although only one set is actually useful. The correlations of x with x and y with y (e.g. female with female in the example) unsurprisingly report a perfect correlation of '1.000'. The important section is the correlation of x with y (female with male in the example). The statistic is given first ($r=0.881$ in the example), then the numbers of pairs of observations in brackets and finally a P-value (labelled 'Sig. (2-tailed)' and 0.020 in the example). As r is positive it shows that there is positive size assortment of individuals in pairs in the example (i.e. large females tend to be paired with large males) and the association is strong enough to allow us to reject the null hypothesis and accept the alternative hypothesis.

R Assuming the data have been imported from a text file that looks exactly like the data in the example and then attached, correlation is very easy to achieve with the 'cor()' function:

```
> cor(Female,Male)
[1] 0.8813196
```

This gives the value of the Pearson correlation coefficient for comparisons of the sizes of males and females in pairs. This is a high value and using r^2 as indication, accounts for about 77% of the variation in size:

```
> 100*cor(Female,Male)^2
[1] 77.67243
```

If you want to correlate all pairs of variables in one go that can be done with 'cor()' too:

```
> cor(penguin)
Pair Female Male
Pair 1.0000000 0.4983785 0.7177929
Female 0.4983785 1.0000000 0.8813196
Male 0.7177929 0.8813196 1.0000000
```

Here my data was in 'penguin' and shows all pairwise correlations, twice. In this case only the female/male correlation is of interest.

If you need more output, and a *P*-value, use the function 'cor.test()':

```
> cor.test(Female,Male)

Pearson's product-moment correlation

data: Female and Male
t=3.7303, df=4, p-value=0.02029
alternative hypothesis: true correlation is not equal
 to 0
95 percent confidence interval:
0.2449766 0.9869616
sample estimates:
cor
0.8813196
```

This confirms that the correlation coefficient of 0.88 is indeed significant at $P=0.02$ and that the 95% confidence intervals for the value of *r* range from 0.24 to 0.98, which indicates that it is very likely that there is a positive relationship between male and female size in pairs and therefore evidence of assortative mating.

MINITAB Input the data in two columns using one row for each pair of observations. Label the columns appropriately. From the 'Stat' menu select 'Basic statistics' then 'Correlation…'. Highlight the names of the two variables then press the select button (Fig. 8.1). Click 'OK'. [*Or type* 'corr c1 c2' *at the MTB> prompt in the session window.*]

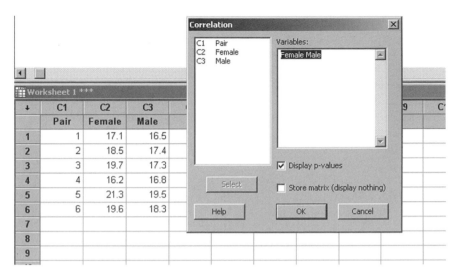

Fig. 8.1 Bivariate correlation in minitab. Each pair of variables should be highlighted and selected to move them into the 'Variables:' box.

The following output appears:

Correlations: Female, Male

```
Pearson correlation of Female and Male = 0.881
P-Value = 0.020
```

The test reports the value of r and the P-value. In the example this indicates that there is a strong positive association between the two variables, showing that there is positive assortative mating. In the example the value of P is well below 0.05, indicating that this is a significant association, even with a very small sample size, and the null hypothesis of no association should be rejected. However, as noted above, it is important to be cautious about the significance of correlations, especially with large samples.

Excel There are two methods for calculating the Pearson's product-moment correlation in Excel: the first requires the Analysis ToolPak to be installed, the second does not. To install the Analysis ToolPak go to the 'Office button', then click the 'Excel Options' button. Choose the 'Add-Ins' menu and here you will be able to add or remove 'Application Add-ins'.

Method 1 Arrange the data in two columns with appropriate column labels. From the 'Data' menu/ribbon select 'Data analysis…' then 'Correlation'. Click 'OK'. The 'Input Range:' box should contain the cell code for the top left and bottom right of the area containing the data separated by a colon (':'). This can be done most easily by selecting the 'Input Range:' box then clicking on the top left cell and dragging to the bottom right. Data in the example are arranged in columns, and that option should be selected by default. If the variable labels are included in the selected cells, the 'Labels in first row' box should be checked, it is a good idea to include the labels as it makes the output easier to understand. The output can either appear on a separate sheet or elsewhere on the same sheet. If on the same sheet the 'Output range' option needs to be selected and the cell indicating the point at the top left of the output put into the box. Click 'OK'.

This table appears:

	A	B	C
1		Female	Male
2	Female	1	
3	Male	0.88132	1

Method 2 Arrange the data in two columns (or two rows). Select an empty cell then click on 'Insert Function', then 'Statistical' and select 'Correl'. Click 'OK'. Input the range of cells containing the data for the first variable into the 'array1' box (this can be done by either typing in the first and last cell with ':' between them or by clicking on the first cell and dragging the pointer to the last cell). Put the range for the second variable in the 'array2' box and click 'OK'. The *r* value appears in the cell.

Either method will produce the same number although the output from method 1 appears to be more extensive. The value of the statistic *r* is given. It is 0.88132 using the example data, indicating strong positive association of the two sets of observations. The r^2 value can be determined by selecting an empty cell and typing '=0.88132^2' (the ^; symbol is often used to mean 'to the power of'). If the *P*-value associated with this value of *r* is required this can be determined using regression (regress one variable on the other and inspect the *P*-value reported in the ANOVA table section of the output; see the regression section later in this chapter for further details). Be warned that *P*-values in correlation are highly influenced by sample size and large samples will give low *P*-values even when the effect, as measured by r^2, is quite weak.

Spearman's rank-order correlation

This is one of two commonly used non-parametric equivalents of the Pearson's product-moment correlation. The statistic it gives is called r_s and ranges from −1 through 0 to 1, indicating perfect negative correlation, no correlation and perfect positive correlation, respectively. Although this is superficially the same scale as the Pearson *r* it is not advisable to compare the results from this test directly with values of *r*. As long as there are two observations for each individual and that the observations are measured on a scale that can be put into a meaningful rank order this test is appropriate. Spearman's correlation is much more conservative than Pearson's.

An example

The same penguin example will be used as for the Pearson product-moment correlation used above. It is quite possible that a researcher will use the Spearman's rank-order correlation rather than the Pearson if there is a doubt about the data set's appropriateness for the Pearson test.

SPSS Arrange the data in two columns with each row representing an individual (in the example the 'individuals' being considered are actually pairs of birds). Label the columns appropriately. From the 'Analyze' menu select 'Correlate' then 'Bivariate…'. Highlight the names of the two variables and move them into the 'Variables:' box. Make sure that 'Spearman' is checked (Pearson is checked by default) and that 'Test of Significance' is 'Two-tailed'. Click 'OK'.

The following output appears:

Nonparametric Correlations

Correlations

			Female	Male
Spearman's rho	Female	Correlation Coefficient	1.000	.771
		Sig. (2-tailed)		.072
		N	6	6
	Male	Correlation Coefficient	.771	1.000
		Sig. (2-tailed)	.072	
		N	6	6

This confirms the test used and then gives the value of r_s (labelled 'Correlation Coefficient'), the number of pairs of observations and the *P*-value associated with the r_s value (labelled 'Sig. (2-tailed)'). In the example although the statistic is clearly showing a large positive association it is not large enough to be deemed significant because the sample size is rather small. However, the *P*-value is sufficiently close to 0.05 to warrant further investigation.

R The data should be arranged as in the example and imported to R. The same functions can be used as for Pearson correlation ('cor()' and 'cor.test()') but with the option 'method= "spearman"' or 'method= "s"' added:

```
> cor(Female,Male,method="spearman")
[1] 0.7714286

> cor.test(Female,Male,method="s")

Spearman's rank correlation rho

data: Female and Male
S=8, p-value=0.1028
```

```
alternative hypothesis: true rho is not equal to 0
sample estimates:
rho
0.7714286
```

The output from the Spearman's correlation is usually called Spearman's *rho*, Spearman's ρ or r_s. Here the value is 0.77, which is quite a strong positive correlation, but the significance is well above 0.05 because the data set is so small.

Note that Spearman's correlation is effectively a Pearson correlation on the ranks which can be demonstrated using the 'rank()' function within 'cor()' in R:

```
> cor(rank(Female),rank(Male))
[1] 0.7714286
```

MINITAB This test cannot be carried out directly in MINITAB. The data first have to be put into rank order and then a Pearson correlation carried out on the ranked data.

1 Input the data in two columns. Label the columns appropriately. If there are any individuals where one or both of the observations are missing these should be omitted from the data set. From the 'Data' menu select 'Rank…'. Put the name of the first variable into the 'Rank data in:' box and put a new name (e.g. 'RankF') in the 'Store ranks in:' box. Click 'OK'. A new column will appear with the original data replaced with integers starting at '1' for the smallest value. Repeat for the second variable.

2 Carry out a normal Pearson correlation on the *ranked* data. From the 'Stat' menu select 'Basic statistics' then 'Correlation…'. Highlight the two *ranked* variables in the list on the left and click on select to move them into the 'Variables:' box. Click 'OK'.

This output appears:

Correlations: RankF, RankM

```
Pearson correlation of RankF and RankM = 0.771
P-Value = 0.072
```

The test indicates that a Pearson correlation has been carried out (the package does not 'know' that ranked data are being used). The correlation reported is the value of r_s. In this case it shows that there is a strong positive association between male and female size, but the significance of 0.072 is not small enough for us to reject the null hypothesis.

Excel There is no direct way of carrying out this test in Excel, even with the Analysis ToolPak add-in installed. However, it is possible to carry out the test by first ranking the data and then performing a normal correlation.

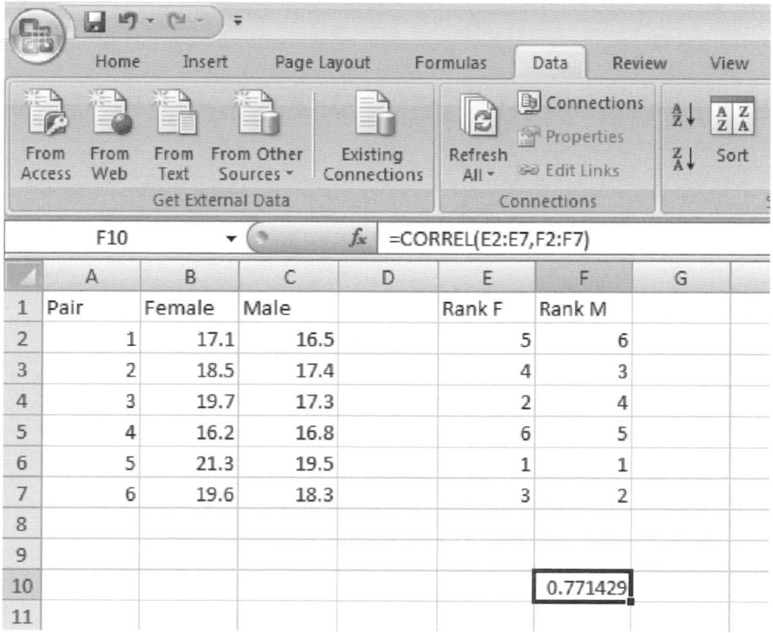

	A	B	C	D	E	F	G
1	Pair	Female	Male		Rank F	Rank M	
2	1	17.1	16.5		5	6	
3	2	18.5	17.4		4	3	
4	3	19.7	17.3		2	4	
5	4	16.2	16.8		6	5	
6	5	21.3	19.5		1	1	
7	6	19.6	18.3		3	2	
8							
9							
10						0.771429	
11							

Fig. 8.2 Spearman's rank correlation in Excel. The data have to be ranked first and then a normal Pearson's product-moment correlation is carried out on the ranked data.

1 Input the data in two columns with a pair of observations on each row. Do not include any rows where one or both of the observations are missing. Use the cells at the top of the columns for appropriate labels. To rank the data first select a cell in an empty column in the same row as the first row of data. To use the 'rank' operation it is probably easiest to simply type '=RANK(B2,B$2:B$7)' rather than use 'Insert function'. This command asks that the cell is given the rank of cell B2 in the range of cells B2 to B7 (the six cells containing the female sizes in the example). The inclusion of the $ symbol makes it very easy to use the copy and paste commands to rank the whole data set as it anchors the row numbers, thus preserving the correct range during the pasting. First select the cell, then click on the small black square in the bottom right and drag to the cell in the next column to the right level with the *last* observation in the data set. This will place the appropriate 'RANK' syntax in each cell without having to type in anything else (in the example it left '=RANK(C7,C$2:C$7)' in the bottom right cell). Label the columns containing the ranks appropriately.

2 Carry out a Pearson correlation on the *ranked* data. Select an empty cell. Click on 'Paste function' and select 'CORREL' from the 'Statistical' section of the list. Click 'OK'. Input the range of cells containing the data for the first variable into the 'Array1' box (this can be done by clicking on the first cell and dragging the pointer to the last cell). Put the range for the second variable in the 'Array2' box and click 'OK' (Fig. 8.2). The r_s value appears in the cell.

In the example the value of r_s is 0.771429, indicating a strong positive association of ranked female and ranked male size. No value of significance is given.

Kendall rank-order correlation

A second, and slightly less widely used, non-parametric correlation in the Kendall rank-order correlation. It is sometimes called Kendall's tau and the statistic produced is usually denoted as T. As with the Spearman's correlation, the test can be carried out on any data set where there are two observations for each individual and the data can be put into a meaningful rank order. Like other measures of association, T ranges from −1 through 0 to 1, indicating the range from perfect negative correlation, to no correlation and perfect positive correlation. Although this is the same as the Pearson r and the Spearman r_s it is not advisable to compare the results from this test directly with either.

The only slight advantage of Kendall correlation over Spearman's is that T can be used in partial correlation whereas r_s cannot.

> It is advisable to use caution when interpreting the significance of a value of T as when sample sizes are very large there is likely to be a P-value less than 0.05 even though the association is only slight.

An example

The same example as used for the Pearson and Spearman tests is used.

SPSS As for Spearman's correlation, arrange the data in two columns with a row for each 'individual'. Label the columns appropriately. From the 'Analyze' menu select 'Correlate' then 'Bivariate…'. Select the two variables and move them into the 'Variables:' box. Select the 'Kendall's tau-b' option from the 'Correlation coefficients:' list. Make sure 'Test of Significance' highlights 'Two-tailed'. Click 'OK' to run the test.

The following output appears:

Nonparametric Correlations

Correlations

			Female	Male
Kendall's tau_b	Female	Correlation Coefficient	1.000	.600
		Sig. (2-tailed)		.091
		N	6	6
	Male	Correlation Coefficient	.600	1.000
		Sig. (2-tailed)	.091	
		N	6	6

This confirms the test used and then gives the value of T (here labelled 'Correlation Coefficient), the number of pairs of observations and the P-value associated with the T value (labelled 'Sig. (2-tailed)'). In the example, although the statistic is clearly showing a large positive association it is not large enough to be deemed significant because the sample size is rather small. However, the P-value is close enough to 0.05 to be worthy of further investigation.

Note that the value of T is rather less than the value of r_s in the Spearman's test although both used the same data. This highlights the fact that the two statistics should not be compared directly.

R The data should be arranged as in the example and imported to R. The same functions can be used as for Pearson correlation ('cor()' and 'cor.test()') but with the option 'method="kendall"' or 'method="k"' added:

```
> cor(Female,Male,method="k")
[1] 0.6

> cor.test(Female,Male,method="kendall")

Kendall's rank correlation tau

data: Female and Male
T=12, p-value=0.1361
alternative hypothesis: true tau is not equal to 0
sample estimates:
tau
0.6
```

The output from the Kendall rank correlation is usually called Kendall's *tau* or τ. Here the value is 0.6, which is a fairly strong positive correlation, but the significance is well above 0.05 because the data set is so small. Note that although all three correlations use the same range (−1 to +1) they cannot be directly compared.

MINITAB A Kendall rank correlation is not possible in MINITAB: use Spearman's rank correlation instead.

Excel There is no direct way of carrying out this test in Excel: use Spearman's rank-order correlation instead.

Regression

The use of regression usually implies that a prediction of one value is being attempted from another; that is, cause and effect. Regression analysis is considered in the next section. However, as most regression output gives a P-value and an r^2 value, the similarity to the Pearson product-moment correlation is very great. The P-value given in a standard linear regression is the probability that the best-fit slope of the relationship between two variables is actually zero. In a comparison with the Pearson statistic, this translates to the probability that there is no relationship (i.e. $r = 0$). Regression analysis usually considers a second null hypothesis: 'the value of y is zero when x is zero'. This translates to a test of whether the best-fit line through the data set passes through the origin. It is often labelled as a test of the intercept.

The advantage of using regression rather than Pearson's correlation is that the assumption that both variables are distributed normally is lifted. The assumptions are different, although slightly less restrictive. For example, regression assumes that the x ('cause') values should be measured without error, that the variation in the y ('effect') is the same for any value of x, that the y values should be normally distributed at any value of x and, for linear regression, that the relationship between two variables can be described by a straight line. Of these the assumption that variance in y is the same for all values of x is probably the least likely to be true. It is usual for variance in y to increase as the x ('cause') variable increases.

If you decide to use regression to determine the association between two variables please use great caution because the implication is that one of the variables in some way depends on the other. Also, one of the underlying assumptions of regression is that the values of the 'cause' variable are in some way set, or chosen, by the investigator; clearly this is not the case if the observations are taken at random.

An example

Again we consider the penguin pairs used as the example throughout this section. The researchers were testing the null hypothesis that male and female birds were forming pairs independent of their size. The alternative hypothesis was that there was an association (either positive or negative) of male and female sizes in pairs. Framed in this way the hypothesis is not suitable for regression. But, if the penguins form pairs by a choice of one sex for another then it might become a more like a regression problem. If females actively chose males then the null hypothesis can be framed in regression terms as 'male size does not depend on female size'. However, even if male size can be said to *depend* on female size there is still the problem that the female sizes were not set or chosen by the investigator.

Is there a cause-and-effect relationship between two variables?

Questions

There are many circumstances where it is clear that one set of observations in some way depends on another. In this section there will be two observations for each 'individual' with one observation being the 'cause', x, 'predictor' or 'independent' variable that is set, or chosen, by the experimenter and the other being the 'effect', y or 'dependent' variable, which is never set by the experimenter. There are a variety of methods that can be applied to determine the form and strength of the relationship between the cause and effect that make different assumptions about the variables and the form of the relationship between them. Five tests are considered here: **linear regression, Kendall robust line-fit method, logistic regression, model II regression** and **polynomial regression**.

'Standard' linear regression

'Standard' linear regression (a.k.a. model I linear regression) is a very widely used statistic in biology. It is also, possibly, the most abused statistic in biology because the assumptions of the test are often flouted. Linear regression is an extremely powerful and useful technique that determines the form and strength of a relationship between two variables. It is used if the intent is to be able to predict a value for y (effect, response or dependent) from a given value of x (cause, predictor or independent). There are several components to any output. The 'slope' is the slope of a straight line of best fit drawn through the set of points with co-ordinates defined by the two variables. Slope can be positive or negative indicating an increase or decrease of y with increasing values of x. The slope can, theoretically, take any value. A slope of zero indicates no change in y with x. The second component of the output is the 'intercept' or 'constant'. This is the predicted value of y when x is equal to zero. The slope is often called b or m and the intercept a or c or 'constant'. Both slope and intercept are usually quoted with some measure of their variability (e.g. a 95% confidence interval or standard deviation). There is often a significance test result given with regression output. This is a test of whether the slope is zero or not (i.e. testing a null hypothesis that $b=0$). If the P-value is less than 0.05 this should be interpreted as an indication that the slope is significantly different from zero, indicating that there is a relationship between the x and y variables.

Linear regression makes many assumptions about the data sets. Important assumptions include that the values of x are measured without error, that the values of x are chosen or set by the experimenter, that the relationship between x and y is best fitted by a straight line ($y=a+bx$), that the variation in y is the same for all values of x and that y is normally distributed for any value of x.

> Tip: if you are unsure which of the two variables is the x ('cause') variable then linear regression is almost certainly not appropriate.

Prediction

Once a best fit-line has been determined then a value for the 'effect' can be predicted for any value of the 'cause'. In practice it is unwise to use values of the 'cause' variable that are beyond the range of the data used to fit the line as the shape of the relationship is unlikely to be the same across all values of the 'cause'. Although it may appear tempting, you should *never* use an observation of the 'effect' to make a prediction of the 'cause' by simple rearrangement of the algebra of the best-fit line.

Interpreting r^2

One commonly used output from linear regression is the value of r^2. This is often expressed as a percentage and described as the amount of variation explained by the regression (i.e. how much of the variation in the 'effect' can be accounted for by using the relationship between 'cause' and 'effect'). Another way to interpret r^2 is to consider that a prediction of the 'effect' can be made from its overall mean and the percentage improvement in prediction in the 'effect' variable that results from the regression is given by r^2.

Comparison of regression and correlation

The assumptions of linear regression are very different to those of correlation. 'Standard' correlation assumes that both x and y are normally distributed. For this reason it is tempting to use regression *in lieu* of correlation when variables are not normally distributed. Do not do this: try an alternative correlation instead because regression makes different assumptions that are unlikely to be true if correlation was the preferred statistical technique.

Residuals

The variation in y not accounted for by the best-fit line relationship between x and y is called the residual variation. For each observation of x there is a predicted value on the line of y. Because y varies, any point is unlikely to lie exactly on the fitted line. The *vertical* distance from the point to the line is the *residual* for that point. It is often helpful to examine the residuals by plotting them against x. This is offered as an option in most statistical packages. If the relationship of x and y is really a straight line, or there is no relationship at all, then the residuals will be scattered without pattern for all values of x. However, if the relationship is really a curve then the residuals of a best-fit straight line will show this to be the case: most of the residuals will be negative (or positive) at the ends of the line and positive (or negative) in the middle. If this is the case, then either a polynomial regression might be a better option than a linear regression or the data should be transformed.

Also, one of the assumptions of regression is that there is the same variation in y for all values of x. If the residuals are all small at one end of the range of x values and large at the other this would indicate that this assumption had been violated: this situation is another common one in biology.

Confidence intervals

The confidence intervals (CIs) attached to slope and intercept allow a range of possible lines to be drawn with boundaries enclosing the range within which 95% (or 90%, if preferred) of the lines of best fit will appear. As this range of possible lines is, in some way, anchored to the y-axis at the point where x is zero and comprises a range of lines of different slope, the 95% confidence limits of the lines are *not* straight lines. They are always curved lines that are closer to the best-fit line in the middle of the range of values of x and further away from it at the extremes. This indicates that you can be more confident about the predictions of the regression in the middle of the data.

Prediction interval

This is a range within which 95% of the values of y are predicted to occur for any given value of x. The 95% prediction intervals (PIs) will always be further away from the best-fit line than the 95% confidence interval for the line. They do not run parallel to the best-fit line but, like the confidence interval for the line, are narrower for mid-range values of x.

An example

A team of researchers is investigating the uptake of an experimental drug through the stomach. They suspect that the acidity of the stomach will affect uptake and propose an experiment to determine the uptake across a range of acidities. Using preparations of sheep stomach and a fixed concentration of the drug in solution, a range of pH values is prepared and the passage of the drug through the stomach monitored. This is suitable for regression as the acidities ('cause') are set by the experimenters and we can assume that they are measured without error.

	pH					
	0.6	0.8	1.0	1.2	1.4	1.6
Uptakes	11.32	11.29	11.37	11.32	11.32	11.49
	11.31	11.22	11.40	11.31	11.36	11.52
	11.22	11.18	11.38	11.35	11.40	11.38
	11.23	11.21	11.37	11.32	11.35	11.49

It is usually best to plot the data as a scattergraph at this point to get a feel for the form of the relationship. This has been done in Excel (Fig. 8.3).

SPSS There are many options that can be applied to enhance the output of the standard linear regression in SPSS. First the very minimum output will be considered and then options that might be useful to help interpret the form of the relationship between the two variables.

Input the data in two columns and label them appropriately. There should be one row for each observation (two numbers). From the 'Analyze' menu select 'Regression' and 'Linear…'. Move the column containing the y, 'response' or 'effect' variable into the 'Dependent:' box. Move the x, 'predictor' or 'cause'

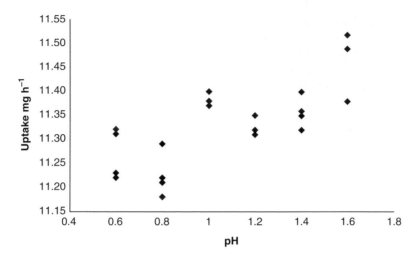

Fig. 8.3 Visualization of data suitable for regression using Excel. There were six pH levels set by the experimenter and four observations of uptake made at each level. This sort of plot gives a feel for the variation and trend in the data before any regression analysis is undertaken. It is right to plot the data like this before a trend line is added, as the line will draw the eye.

variable (the one set or chosen) into the 'Independent(s):' box. Ignore all of the possible options, just click 'OK'.

Using the example data the following output appears:

Regression

Variables Entered/Removed[b]

Model	Variables Entered	Variables Removed	Method
1	pH[a]		Enter

a. All requested variables entered.

b. Dependent Variable: Uptake.

Model Summary

Model	R	R Square	Adjusted R Square	Std. Error of the Estimate
1	.758[a]	.574	.555	.05893

a. Predictors: (Constant), pH

ANOVA[b]

Model		Sum of Squares	df	Mean Square	F	Sig.
1	Regression	.103	1	.103	29.654	.000[a]
	Residual	.076	22	.003		
	Total	.179	23			

a. Predictors: (Constant), pH

b. Dependent Variable: Uptake

And

Coefficients[a]

Model		Unstandardized Coefficients		Standardized Coefficients	t	Sig.
		B	Std. Error	Beta		
1	(Constant)	11.127	.041		274.296	.000
	pH	.192	.035	.758	5.446	.000

a. Dependent Variable: Uptake

Even this minimum level of output is rather bewildering. First the output confirms the test used and then the names of the x variables (just one in most cases). Then comes a 'Model Summary' that gives the value of r (labelled 'R') and r^2 (labelled 'R Square'). The other columns give a more conservative estimate of r^2 and a measure of the confidence in r^2.

Next is an ANOVA table comparing the best-fit line against a null hypothesis of no relationship (e.g. slope is zero). In this table the F-ratio ('F') and its associated P-value (labelled 'Sig.' here) is given. In the example the P-value is given as '.000' meaning $P<0.001$, which is highly significant. There is very little doubt that a significant amount of the variation in y is explained by x.

Finally come the estimates for the values of the two 'coefficients' (i.e. the slope and the intercept). The intercept is given first on a row (labelled '(Constant)') here giving a value with standard error and then a t-test of the null hypothesis that the intercept is zero. The next line (labelled 'pH' in the example) gives a value for the slope, and is followed by an estimate of the standard error associated with the slope and then a t-test of the null hypothesis that the slope is equal to zero (this is essentially a repeat of the result of the ANOVA table).

In this example the slope of the relationship between 'pH' and 'uptake' is 0.192 and the intercept (value of 'uptake' when 'pH' is zero) is 11.127. Both these values are significantly different from zero. The best-fit line of 'uptake$=0.192\times$pH$+11.127$' explains about 57% of the variation (the r^2 value) in the observations of 'uptake'. The best-fit line can be, theoretically, applied to any value of x ('pH'), within the range used in the analysis, to make a prediction of y ('uptake').

Useful options to help with the interpretation of the regression relationship include a plot of the residuals. From the 'Linear Regression' dialogue box choose 'Plots'. Then put the 'DEPENDNT' into the 'X:' box and ''ZRESID' into the 'Y:' box. This will generate a plot of the residuals plotted against the values of the 'effect' variable. If there is a clear pattern in this plot then it is likely that one of the assumptions of the regression has been violated. For example, if the true relationship is a curve this will show up in this plot as a curved shape.

R Executing a simple linear regression in R is very simple in the 'lm()' function. Assuming that the data have been attached and are two labelled columns, give the function the name of the response variable, then a tilde (~) then the name of the predictor variable. Use the 'summary()' function to provide useful output:

```
> summary(lm(Uptake~pH))

Call:

lm(formula=Uptake ~ pH)

Residuals:
Min          1Q          Median      3Q          Max
-0.10038    -0.04586    -0.00956    0.05619     0.08619

Coefficients:
Estimate Std. Error t value Pr(>|t|)
(Intercept)    11.12695   0.04057    274.296    < 2e-16   ***
pH             0.19179    0.03522    5.446      1.80e-05  ***
---
Signif. codes: 0 '***' 0.001 '**' 0.01 '*' 0.05 '.' 0.1 ' ' 1

Residual standard error: 0.05893 on 22 degrees of
   freedom
Multiple R-squared: 0.5741, Adjusted R-squared: 0.5547
F-statistic: 29.65 on 1 and 22 DF, p-value: 1.804e-05
```

The output gives an idea of the model that is being used in the 'lm()' function, then gives a report of the distribution of the residuals; that is, the vertical distance of each point from the fitted line. We hope that the residuals are symmetrical and have a median of zero as appears to be the case here. Next comes the important estimates of the parameters of the line and the significance of those parameters. First is the intercept: this is the value of y when x is zero (or in the example the value of 'Uptake' when 'pH' is zero), this is 11.13 here and highly significantly different from zero. However, this significance is of little interest, as it is in most regression analyses. The second line is the more important one.

It gives the value of the slope of the relationship, here 0.19, and provides a significance value, here a highly significant 0.000018. This means in the example that for each increase of 1 in 'pH' we would predict an increase of 0.19 in 'Uptake' and this slope is very significantly different from the null hypothesis that the slope is zero.

The r^2 value is 0.574. This indicates that 57% of the variation in the response variable ('Uptake') can be explained by the predictor variable ('pH'). This can be visualized, as in Fig. 8.3:

```
> plot(pH,Uptake)
```

A best-fit line can be added to this plot by taking the intercept and slope values from the output and using 'abline()':

```
> abline(11.127,0.192)
```

A series of plots of the data and residuals can be produced:

```
> m<-lm(Uptake~pH)
> plot(m)
```

MINITAB Put the data in two columns: one for the x values and one for the y values. Label the columns appropriately. From the 'Stat' menu select 'Regression' then 'Regression…'. In the dialogue box that appears move the column containing the y values into the 'Response:' box and the column of x values into the 'Predictors:' box. Ignore the array of additional options for the moment. Click 'OK'. Note that MINITAB calls the y or 'effect' variable the 'response' and the x or 'cause' variable the 'predictor'.

(Or, if the command interface is enabled, and assuming the y values are in c1 and the x values in c2, type 'Regress c1 1 c2' at the MTB> prompt in the session window.)

This output appears:

```
Regression Analysis: Uptake versus pH

The regression equation is
Uptake = 11.1 + 0.192 pH

Predictor     Coef    SE Coef       T          P
Constant   11.1270    0.0406   274.30      0.000
pH          0.19179   0.03522     5.45      0.000

S = 0.0589325 R-Sq = 57.4% R-Sq(adj) = 55.5%

Analysis of Variance

Source          DF        SS       MS        F         P
Regression       1   0.10299  0.10299    29.65     0.000
Residual Error  22   0.07641  0.00347
Total           23   0.17940
```

The output first confirms the test and then gives the best-fit line of the two variables. This describes both the intercept (11.1 in the example) and the slope (0.192 in the example). Then comes a section repeating the two parameters where the intercept is labelled as 'Constant'. A standard deviation is given for each along with a *t*-test with the null hypothesis that the parameters are zero and the *P*-value (labelled 'P') associated with the *t*-test (labelled 'T'). In the example both slope and intercept are highly significantly different from zero, although you will rarely be interested in whether the intercept is significantly different from zero. Next comes a line that reports the proportion of the variance in the *y* values that is explained by the best-fit line. In this case the r^2 value (labelled 'R-sq') is 57.4%.

Finally comes an ANOVA table that compares the variation explained by the best-fit line with the residual variation. The null hypothesis is that none of the variation in *y* is explained by the regression line. In this case the relationship is highly significantly different from zero and *P* is given as '0.000' although this should be reported as $P < 0.001$.

There are many optional extras that can be added to the basic regression analysis. Some of these extras are valuable for exploring the relationship between the *x* and *y* variables. The plot of 'Residuals versus fits' can often be revealing and is a good way to detect whether the linear relationship would be better fitted to a curve.

Excel Regression is only possible if the Analysis ToolPak has been installed. Put the data into two columns. One for the 'effect' and one for the 'cause'. Each row should represent a single observation (i.e. two values). Use the cells above the data to label the columns appropriately. From the 'Data' menu/ribbon select 'Data Analysis' then select 'Regression' from the long list and click 'OK'.

The 'Input Y Range:' box should contain the code for the first and last cells with the list of *y* values (including the label). This can be most easily done if you click in the box to move the cursor there then click on the top of the list of data and drag to the bottom of the list. Repeat this for the 'Input X Range:' box with the *x* values (these are the ones set or chosen by the experimenter). If the labels have been included then make sure that the 'Labels' option is checked. Note that Excel asks for the *y* ('effect') values first, although it might be easier for graph drawing to have the *x* values to the left of the *y* values.

To define where the output is to appear choose either 'Output range:' and point at the top left corner of where the output will appear or leave the selection as 'New Worksheet Ply:', which will put the output in a fresh worksheet. Click 'OK'.

The example data produces this output:

SUMMARY OUTPUT

Regression Statistics	
Multiple R	0.757685706
R Square	0.574087629
Adjusted R Square	0.554727976
Standard Error	0.058932513
Observations	24

ANOVA

	df	SS	MS	F	Significance F
Regression	1	0.102988929	0.102988929	29.65381775	1.80402E–05
Residual	22	0.076406905	0.003473041		
Total	23	0.179395833			

	Coefficients	Standard Error	t Stat	P-value	Lower 95%	Upper 95%	Lower 95.0%	Upper 95.0%
Intercept	11.12695238	0.040565502	274.2959362	2.24017E-40	11.04282468	11.21108	11.04282	11.21108
pH	0.191785714	0.035218913	5.445531907	1.80402E-05	0.11874616	0.264825	0.118746	0.264825

This output is rather confusing. First the output confirms that 'Regression statistics' are being reported. Then come four lines giving the strength of the relationship between y and x. This is effectively the r of the Pearson's correlation. The 'R Square' (usually given as 'r-square') can be interpreted as the proportion of the variation in y explained by the best-fit line. In the example just over 57% of the variation in y is explained by the line (0.57408).

Next is an ANOVA table comparing the best-fit line against a null hypothesis of no relationship (i.e. the slope is zero). In the example the P-value (labelled as 'Significance F') is given as '1.80402E-05' meaning 0.000018 or $P < 0.0001$, which is highly significant. There is very little doubt that a significant amount of the variation in y is explained by x.

Finally come the estimates for the two parameters in a separate table (i.e. the slope and the intercept). The first row refers to the intercept giving a value with standard error and then a t-test of the null hypothesis that the intercept is zero with associated P-value (ludicrously small in the example) and then estimates of the 95% confidence intervals for the intercept (in the example the intercept is 95% likely to lie between 11.04 and 11.21). The slope is given on a row labelled 'pH' in the example. The estimated value for the slope is given, and is followed by an estimate of the standard error associated with the slope and then a *t*-test of the null hypothesis that the slope is equal to zero with *P*-value (this is essentially a repeat of the result of the ANOVA table) followed by 95% confidence intervals for the slope.

In this example the slope of the relationship between 'pH' and 'uptake' is 0.1918 and the intercept (value of 'uptake' when 'pH' is zero) is 11.13. Both these values are significantly different from zero. The results can be interpreted as a best-fit line of 'uptake = 0.1918 × pH + 11.13' explaining 57% of the variation in the observations of 'uptake'. This best-fit line could be, theoretically, applied to any value of 'pH' to predict 'uptake'.

Options to try in the Excel regression dialogue box include the residual plot and table of residuals, where the residual value (distance from the best-fit line) is given for every data point. The residual plot can be used to detect systematic deviations from the assumptions of a linear regression: for example, larger deviations at higher values of x, or if the true line is a curve.

Kendall robust line-fit method

This is a simple non-parametric test that can be used instead of normal regression. It is unlikely to be supported by a statistical package although it should be possible to calculate using simple spreadsheet manipulations. The idea is to calculate the slope of the line between every possible pair of x, y points. This will potentially generate a large number of slopes (e.g. ten slopes from five points, or 45 slopes from 10 points). The median slope is then selected as the best estimate of b.

Once the median slope has been determined the intercept can be calculated by taking the slope back from every observation to $x = 0$. This gives an intercept for every point. The median intercept value should be used.

This test makes very few assumptions about the data other than it is measured on a meaningful scale.

Logistic regression

This is a special form of regression analysis that is used when the 'dependent' or 'effect' variable can only be classified into groups (many packages limit this to two groups). It is a regression analysis that uses the proportions of the two possibilities of the 'dependent' converted to a logit to calculate the relationship with the 'independent' or 'cause' variable and is particularly useful when there are proportions near 0 or 1.

Unlike linear regression, the logistic regression does not require the 'cause' ('independent' or x) to be on a meaningful scale. There can be as few as two points on the x-axis (e.g. male and female) and a logistic regression is still possible. Logistic regression uses logit transformations (see Chapter 5) of the 'effect' ('dependent' or y) values although the packages make this transformation for you.

Logistic regression is a very powerful technique and can be used as a classification tool in a similar way to discriminant function analysis. The example below is a very simple problem for logistic regression. The problem is a similar one to a linear regression but is using a 'response' or 'effect' variable that can only be in one of two categories. Logistic regression can also be used as a form of multiple regression with many 'cause' or 'predictor' variables.

An example

The null hypothesis is that the prevalence of a plant virus is unaffected by shading conditions. Researchers were able to estimate canopy cover and, by using a light meter, identified seven shading levels. Within each shade level 10 plants were selected at random and the presence or absence of the virus on each plant recorded. The results are tabulated below.

Shade category	1	2	3	4	5	6	7
Plants with virus	2	4	4	4	6	6	8
Plants without virus	8	6	6	6	4	4	2

SPSS The data should be entered as 70 rows of data with one row for each individual. There should be two columns one for shade category and one for virus coded at either 0 or 1. From the 'Analyze' menu, select 'Regression' then 'Binary Logistic…'. In the dialogue box more the 'effect' variable ('Virus' in the example) to the 'Dependent:' box. Move the 'cause' variable to the 'Covariates:' box. Click 'OK' to continue.

Masses of output is produced. Here is the most important bit:

Classification Table[a]

Observed		Predicted		
		virus		Percentage Correct
		.00	1.00	
Step 1	virus .00	26	10	72.2
	1.00	14	20	58.8
	Overall Percentage			65.7

a. The cut value is .500

Variables in the Equation

		B	S.E.	Wald	df	Sig.	Exp(B)
Step 1[a]	shade	.370	.134	7.634	1	.006	1.447
	Constant	−1.544	.598	6.672	1	.010	.214

a. Variable(s) entered on step 1: shade.

The first table shows how many of the observations are correctly assigned to category using the model generated by the logistic regression. In the example of the 34 plants without virus there were 26 (or 72%) correctly assigned to that category by the model. The second table is very similar to a standard regression table. The 'shade' line is the test of the 'slope' and can be interpreted as a test of whether accounting for shade reduces the number of observations misclassified. In the example this has a *P*-value ('Sig.') of 0.006, indicating that the null hypothesis should be rejected and that shade does indeed have a very significant effect on the presence of virus. The second line is a test of the 'intercept' and is only useful if you are interested in the proportion of plants infected when shade category is zero (i.e. not relevant in this case).

R The data can be in 70 rows as for SPSS and MINITAB, or as the proportion with virus for each shade category. If the data are in 70 rows and two columns, shade category and virus, which has been coded as true/false, then:

```
Shade       Virus
1           T
2           F
3           T
  ...
```

For logistic regression the 'glm()' function is required with the errors set as binomial (i.e. two options such as true/false or yes/no). The response variable is given first followed by a tilde and then the predictor variable. The 'summary()' function, as usual, organizes the output in a usable way:

```
> summary(glm(Virus~Shade,binomial))

Call:
glm(formula=Virus ~ Shade, family=binomial)

Deviance Residuals:
Min         1Q          Median      3Q          Max
-1.6406     -0.9993     -0.7340     1.0510      1.6990

Coefficients:
              Estimate  Std. Error  z value   Pr(>|z|)
(Intercept)  -1.5436    0.5976      -2.583    0.00979 **
Shade         0.3697    0.1338       2.763    0.00573 **
---
```

```
Signif. codes: 0 '***' 0.001 '**' 0.01 '*' 0.05 '.' 0.1 ' ' 1

(Dispersion parameter for binomial family taken to
be 1)

Null deviance: 96.983 on 69 degrees of freedom
Residual deviance: 88.442 on 68 degrees of freedom
AIC: 92.442

Number of Fisher Scoring iterations: 4
```

R provides confirmation of the test and the model that is being used then gives a summary of the deviations from the best-fit model. Then gives a summary of the important results. The line for intercept is only interesting if you want to know whether there are no plants with the virus in shade category zero. The 'shade' line is the 'slope' of the line and has a *P*-value well under 0.01 indicating that shade has a strong influence on the probability that a plant has the virus. Next comes a coding for the number of asterisks, '**' indicates a *P*-value between 0.001 and 0.01.

Then comes a comment about the shape of the assumed distribution of errors (this can be modified) and then a comparison of the deviance of the residuals. If the distribution of errors follows that specified then the residual deviance value will be the same as the degrees of freedom. Here the residuals are larger than the degrees of freedom which indicates the data are more dispersed than is being accounted for. After this comes a value for 'AIC'. This is the Akaike Information Criterion which is a way of accounting for the number of parameters used in a model when comparing the ability of models to fit the data.

This will provide a visual fit of the line:

```
> lgt<-glm(Virus~Shade,binomial)
> plot(Shade, fitted(lgt))
```

MINITAB The data can be entered as 70 rows of data with one row for each individual. There should be two columns one for shade category and one for virus coded at either 0 or 1. Alternatively the data are entered as a frequency table. For this example I assume one row per individual.

From the 'Stat' menu, select 'Regression' then 'Binary Logistic Regression…'. Move the 'effect' variable ('Virus' in the example) to the 'Response:' box and the 'cause' variable ('Shade' in the example) to the 'Model:' box. Click 'OK'.

Here is a portion of the output that appears:

Binary Logistic Regression: Virus versus Shade

```
Link Function: Logit

Response Information

Variable   Value      Count
Virus      T            34      (Event)
           F            36
           Total        70

Logistic Regression Table
                                                   Odds      95%       CI
Predictor        Coef     SE Coef       Z       P  Ratio   Lower     Upper
Constant     -1.54361    0.597575   -2.58   0.010
Shade         0.369703   0.133806    2.76   0.006   1.45    1.11      1.88

Log-Likelihood = -44.221
Test that all slopes are zero: G = 8.541, DF = 1, P-Value = 0.003

Goodness-of-Fit Tests

Method          Chi-Square      DF              P
Pearson          1.13052         5          0.951
Deviance         1.12489         5          0.952
Hosmer-Lemeshow  1.13052         5          0.951

Table of Observed and Expected Frequencies:
(See Hosmer-Lemeshow Test for the Pearson Chi-Square Statistic)

                            Group
Value     1      2      3      4      5      6      7     Total
T
  Obs     2      4      4      4      6      6      8      34
  Exp    2.4    3.1    3.9    4.8    5.8    6.6    7.4
F
  Obs     8      6      6      6      4      4      2      36
  Exp    7.6    6.9    6.1    5.2    4.2    3.4    2.6
Total    10     10     10     10     10     10     10      70
```

The first section gives a frequency table, number of 'true' and 'false' in the example. The next section is very similar to a standard regression table. The 'Shade' line can be interpreted as a test of whether accounting for shade reduces the number of observations misclassified. In the example this has a *P*-value ('Sig.') of 0.006 indicating that shade does indeed have a very significant effect on the presence of virus. The 'Constant' line is a test of the 'intercept' and is only useful if you are interested in the proportion of plants infected when shade category is zero (i.e. not relevant in this case). The next section shows a range of 'goodness of fit' test results and then the observed and expected values generated by the logistic regression for each category of the 'cause' variable. In the example for shade category 4 the model predicted 4.8 to be infected and 5.2 not, with observed values of 4 and 6. Of the 34 plants without virus there were 26 (or 72%) correctly assigned to that category by the model.

Excel There is no direct way of carrying out this test in Excel.

Model II regression

The model two regression is actually a group of analyses that make far fewer assumptions about the data than standard, model I regression. The main culprits are the assumptions in standard regression that the x values are measured without error and that the variation in y (effect) is the same for any level of x (cause). If these assumptions do not hold then model II regression is appropriate. Unfortunately the statistical manipulations required are still being developed and there are several techniques that are not wholly satisfactory. Moreover, they are unlikely to be supported by computer packages.

One suggested model II regression is the Bartlett's three-group method. This method simply arranges the data by the magnitude of the x values and divides into three groups, making sure that there are equal numbers of observations in the largest and smallest thirds. The mean values of x and y are calculated for these two groups only and the slope of the line between the two mean points is calculated.

Another model II regression is the Kendall robust line-fit method described above (page 230).

Polynomial, cubic and quadratic regression

One of the assumptions of standard regression is that the form of the relationship between x and y is a straight line. If this is not true then the first option is to transform either x or y to make the data better fit a straight line. If this does not help then the assumption can be discarded and polynomial regression (of which quadratic regression is a special case) can be employed. The only difference between polynomial and linear regression is that the best-fit line is not straight. The advantage is that a curved line is almost always a better fit, allowing a better prediction of y for each value of x. The disadvantage is that extra parameters have to be included. There is no longer a single 'slope' value but rather two or more factors that have to be applied to x, x^2, x^3 and so on. It is usually either a scatterplot or inspection of the residuals from a linear regression which indicates that the straight line is not the best fit and that some sort of curve should be tried.

It should be noted that each time an extra parameter is added to a regression line the best fit should be better than with fewer parameters. It is possible to get an almost perfect fit to a scatter of points if enough parameters are used. However, the question is: will adding an extra parameter be worth it? This is a judgement call, but your judgement can often be guided by considering the biology. It might be possible to consider that a biological process should be best described by a curved line of some kind, but as the line becomes more complicated the biology underlying the relationship also becomes more difficult to justify.

There are methods available to determine whether adding more parameters is worth it. The Akaike Information Criterion is a commonly used method that accounts for the accuracy of the model fit to the data and the number of parameters used. It can help with model selection when several are suggested.

Tests for more than two variables

Tests of association

Questions

Most test of association assume that there are only two variables being considered. However, it is often the case that for each 'individual' there are three or more observations. If all that is required is to investigate associations in pairs to examine the strength of associations when other variables are accounted for then this section considers briefly **correlation, partial correlation** and its non-parametric equivalent **Kendall partial rank-order correlation**. Once there are several variables for each individual then multivariate analyses considered in the next chapter will often be more appropriate.

Correlation

Simplest method used to analyse data sets with more than two observations for each 'individual' or 'site' is to consider all the possible two-way comparisons that could be made. Statistical packages allow this and will happily produce a large matrix of correlation coefficients that can be trawled to find the largest positive and negative numbers for further investigation.

Note: if the package reports a significance value for each of the correlation coefficients then they should be treated with some caution. Remember that the *P*-value is just the probability of encountering data this extreme or more extreme if the null hypothesis is true and we usually reject the null hypothesis when $P < 0.05$. This implies that 20 correlations will produce, on average, one *P*-value that is less than 0.05 even when there is no association at all. Therefore if there are 10 variables for each 'individual' giving 45 possible pairwise correlation coefficients the chances are very high that one or more will report *P*-values less than 0.05.

There are two commonly used methods for reducing the critical value for *P*. The Bonferroni method simply divides the critical *P*-value by the number of tests carried out. So three tests with an overall critical *P* of 0.05 would give a critical *P* of 0.05/3 for each test. The Dunn–Sidák reduces the critical *P* using the formula $0.95^{1/k}$, where *k* is the number of tests that have been carried out (assuming a critical *P*-value of 0.05).

See the section on correlation for the mechanics of using the packages (pages 210–220).

Partial correlation

A partial correlation coefficient gives a measure of the relationship between two variables when one or more other variables have been held constant. A common use of this technique is in morphological analysis when several variables all relate in some way to size and therefore the bivariate correlation matrix technique employed in the previous section merely confirms that all measures are strongly correlated with each other. The partial correlation of two variables when 'size' is held constant will reveal whether they are related in any other way.

Kendall partial rank-order correlation

The Kendall's rank correlation coefficient is the non-parametric equivalent of Pearson's product-moment correlation coefficient which can be used in partial correlation analysis. If variables are known to violate the assumptions of parametric correlation then this technique should be employed. Unfortunately, this technique is rarely supported in statistical packages.

Cause(s) and effect(s)

Questions

If there are more than two variables which can be labelled 'cause' and 'effect' there are a variety of techniques that can be applied to determine more about the relationship between them. Some of these techniques are simple extensions of ANOVA, regression or correlation analysis while others point towards techniques for data exploration that are covered in the next chapter. **Regression** is used when there are two or more similar 'effect' variables matched to the same 'cause'. **Analysis of covariance (ANCOVA)** when two variables are known to be associated and one is used as the dependent variable in an ANOVA analysis. **Multiple regression** is used when there are several 'cause' variables and a single 'effect'. **Stepwise regression** also has several 'cause' variables and one 'effect' but builds the best-fit model in stages. **Path analysis** is more of a data-exploration technique that arranges the interrelationships between several 'causes' and 'effects'.

Regression

If regression has been carried out on several different sets of individuals, in different sites or in different years, for example, then it is possible to compare the slopes of these different analyses to see if they differ. Most statistical packages do not support this type of analysis directly although a visual comparison of several analyses can be made by plotting the estimated slopes and their confidence intervals to see if they overlap.

Analysis of covariance (ANCOVA)

This technique is something of a hybrid between ANOVA and linear regression. Imagine that a field experiment has been set up with plots given one of five different levels of additional CO_2. Data are collected for sap sugar concentration. This appears to be a simple ANOVA type of design. However, the plots have been surveyed for a range of physical parameters and are known to differ in the amount of organic material in the soil and this affects the sap sugar concentration. The ANCOVA test will effectively use a regression analysis to remove the effect of the organic material level *before* the standard ANOVA is attempted. The factor accounted for by the regression is called the *covariate*, hence the name analysis of covariance. ANCOVA is supported by both MINITAB and SPSS and is easily accessed in R.

ANCOVA can be used with any ANOVA design and is a very useful statistic whenever something known to affect the data can be quantified accurately. ANCOVA makes all the assumptions of ANOVA as well as those of regression. It makes the further assumption that all groups have the same linear relationship with the covariate.

An example A laboratory investigation into the physiological differences between two species of amphipod (shrimp) has shown that although females of both species push water across their developing eggs they do so with different beat frequencies. A further investigation has been made using video cameras in the field to see if this relationship still holds. The investigators also know that beat frequency for both species increases with water temperature and this has been recorded.

Species	Water temperature, °C	Beats/minute
1	10.1	89.0
1	12.2	94.8
1	13.5	99.6
1	11.2	93.8
1	10.2	91.0
1	9.8	89.2
2	14.1	104.5
2	12.3	103.6
2	9.5	91.1
2	11.6	99.6
2	10.1	99.1
2	9.4	88.7

ANOVA or *t*-test shows that the beats/minute is not significantly different between the two species (ANOVA: $F_{1,10} = 2.419$, $P > 0.1$). However, this does not account for the effect of temperature on beat frequency. An ANCOVA using species as the grouping variable and temperature as a covariate shows that there is a highly significant difference between the two species ($F_{1,9}=12.436$, $P=0.006$). Note that ANCOVA has one fewer degrees of freedom for each covariate that is used than an ANOVA on the same data.

SPSS Input the data in columns as in the example. Label the columns appropriately.

From the 'Analyze' menu select 'General Linear Model' then 'Univariate…'. Move the observed data ('Beats' in the example) to the 'Dependent Variable:' box. Move the grouping variable ('Species' in the example) to the 'Fixed Factor(s):' box. Move the covariate ('Temp' in the example) to the 'Covariate(s):' box. Click 'OK' to run the test.

Univariate Analysis of Variance

Between-Subjects Factors

		N
Species	1	6
	2	6

Tests of Between-Subjects Effects

Dependent Variable: Beats

Source	Type III Sum of Squares	df	Mean Square	F	Sig.
Corrected Model	312.262[a]	2	156.131	27.145	.000
Intercept	842.580	1	842.580	146.494	.000
Temp	241.209	1	241.209	41.937	.000
Species	71.053	1	71.053	12.354	.007
Error	51.765	9	5.752		
Total	109425.360	12			
Corrected Total	364.027	11			

a. R Squared = .858 (Adjusted R Squared = .826)

This confirms the number of observations for each level of the fixed factor ('Species' in the example). The rest of the output is an ANOVA table designed to cope with many factors and therefore has extra lines that appear totally superfluous for this simplest possible scenario. The first column, 'Source', gives the source of the variation. It is the line labelled with the name of the main effect that is most important ('Species' in the example). The other columns give: 'Type III sum of squares', then 'df' or degrees of freedom, then 'Mean Square' (the mean square value is the sum of square value divided by the degrees of freedom). As there are two species there is one degree of freedom for 'Species', and the covariate has one degree of freedom. There were six observations within each species giving 11 degrees of freedom for 'Corrected Total'.

Finally comes the important bit; the *F*-ratio, labelled 'F' here. This is the mean square for 'Species' or 'Temp' divided by that for 'Error' (or residual). SPSS gives the *P*-value associated with this value of 'F' and these degrees of freedom and labels it 'Sig.'. In biology we usually look for a value less than 0.05. Here the probability is 0.007 for 'Species' and indicates that the mean 'beats/minute' for the two species are significantly different from each other once the effect of temperature is accounted for.

R ANCOVA is easily reached using the 'aov()' function using the syntax '+ variable name' to indicate that the predictor variable is a covariate. The results can then be displayed using the 'summary()'

```
> summary(aov(BPM~Species+Temp))
            Df  Sum Sq    Mean Sq   F value  Pr(>F)
Species     1   71.053    71.053    12.354   0.0065693  **
Temp        1   241.209   241.209   41.937   0.0001146  ***
Residuals   9   51.765    5.752
---
Signif. codes: 0 `***' 0.001 `**' 0.01 `*' 0.05 `.' 0.1 ` '
1
```

This output gives the basic ANOVA table. The covariate 'Temp' has one degree of freedom and clearly has a highly significant effect on the response variable ($P \ll 0.001$). With the covariate accounted for the factor 'Species' has a significant effect ($P < 0.01$).

You can use the 'lm()' function, to generate the same output:

```
> summary.aov(lm(BPM~Species+Temp))
```

and the 'lm()' function is better than 'aov()' when estimates of the parameters are required:

```
> summary(lm(BPM~Species+Temp))

Call:
lm(formula=BPM ~ Species + Temp)

Residuals:
Min          1Q         Median     3Q         Max
-3.84079    -1.30163    0.07028    0.84091    4.48858

Coefficients:
              Estimate   Std. Error   t value   Pr(>|t|)
(Intercept)   59.8685    5.1938       11.527    1.08e-06   ***
SpeciesB      4.8667     1.3846       3.515     0.006569   **
Temp          2.9580     0.4568       6.476     0.000115   ***
---
```

```
Signif. codes: 0 '***' 0.001 '**' 0.01 '*' 0.05 '.' 0.1
' ' 1

Residual standard error: 2.398 on 9 degrees of freedom
Multiple R-squared: 0.8578, Adjusted R-squared: 0.8262
F-statistic: 27.15 on 2 and 9 DF, p-value: 0.0001542
```

This output shows that the model (factor and covariate) explains over 80% of the variation in the response variable and this is a highly significant fit (*P*=0.0001542).

MINITAB ANCOVA is very simple to carry out in MINITAB. Input the data in columns, exactly as in the example, and label appropriately. From the 'Stat' menu select 'ANOVA' then 'General Linear Model…'. Move the observed data ('BPM' in the example) into the 'Responses:' box, the grouping variable ('Species' in the example) into the 'Model:' box. Click on the 'Covariates…' button and add the covariate ('Temp' in the example) to the 'Covariates:' box. Click 'OK' twice.

   (Or, if the command interface is enabled, type 'GLM c1=c2;' *at the MTB> prompt and* 'covariates c3.' *at the SUBC> prompt, assuming that the data is in c1, the main effect is in c2 and the covariate is in c3. Or you can input commands using* 'Edit' *menu then* 'Command Line Editor'.)

```
General Linear Model: BPM versus Species

Factor      Type   Levels  Values
Species     fixed      2   1, 2

Analysis of Variance for BPM, using Adjusted SS for Tests

Source    DF      Seq SS   Adj SS   Adj MS       F      P
Temp       1      241.21   241.21   241.21   41.94  0.000
Species    1       71.05    71.05    71.05   12.35  0.007
Error      9       51.76    51.76     5.75
Total     11      364.03

S = 2.39826  R-Sq = 85.78%  R-Sq(adj) = 82.62%

Term         Coef   SE Coef       T       P
Constant   62.302     5.147   12.10   0.000
Temp        2.9580    0.4568    6.48   0.000

Unusual Observations for BPM

Obs      BPM      Fit    SE Fit    Residual    St Resid
 11   99.100   94.611     1.094       4.489       2.10 R

R denotes an observation with a large standardized residual.
```

The output first confirms that the test is a general linear model. Then lists the factor(s) (grouping variables) giving the number of groups (labelled 'Levels') and the codes assigned to the groups (labelled 'Values').

   Then comes another confirmation of the test followed by a simple ANOVA table with an extra row for the covariate. The table has degrees of freedom

('DF'), two versions of the sum of squares ('Seq SS' and 'Adj SS') mean square ('Adj MS' which is SS/DF), the F-ratio ('F', which is factor MS/error MS) and finally the P-value ('P'). If the P-value is less than 0.05 then the null hypothesis is rejected. In the example the effect of temperature as a covariate is confirmed to be highly significant 'P' is given as '0.000' which should be reported as $P<0.001$. The factor 'species' has a P-value of '0.007' so the null hypothesis that species have the same 'BPM' is rejected and the alternative hypothesis that the two species differ is accepted. Next comes some of the statistics associated with the regression of the covariate, including a value for r^2 (R-Sq).

Finally there is a regression-style analysis of the covariate ('Temp' in the example). It shows that the relationship has a positive slope ('Coef' can be thought of as the slope of the relationship and is '2.958' in the example). The probability that the slope is zero is very small, given as '0.000' in the example, or $P<0.001$.

The 'Unusual Observations' are observations with very high residuals and they should be looked at more closely as they are likely to be mis-typed or mis-recorded values.

Excel There is no simple way of carrying out ANCOVA in Excel.

Multiple regression

If there are several 'cause' variables set, or chosen, by the experimenter and a single 'effect' variable then multiple regression may be appropriate. The same assumptions as for linear regression apply so each of the 'cause' variables must be measured without error. There is an additional assumption that each of the 'cause' variables must be independent of each other. Multiple regression works in exactly the same way as linear regression only the best-fit line is made up of a separate 'slope' for each of the 'cause' variables. There is still a single 'intercept' which is the predicted value of the 'effect' variable when all the 'cause' variables are zero.

A multiple regression using just two 'predictor' or 'cause' variables can be visualized using a three-dimensional scatterplot, but if there are any more 'predictors' there is no way to satisfactorily display the relationship.

The technique of multiple regression is rather overused as it is supported, and fairly easily accessible, in most statistical packages. There is the implicit assumption that all the 'cause' variables impinge directly on the 'effect' variable that may appear uncomfortable or unreasonable in certain circumstances. Multiple regression can be found in SPSS, R and MINITAB.

Stepwise regression

The assumptions and conditions of stepwise regression are identical to those of multiple regression. The difference is in the way the best-fit model is generated. In stepwise regression the 'causes' are added and subtracted in steps only using those combinations and slopes that generate a better fit (i.e. smaller residuals). This technique is useful as it will help to identify those 'cause' variables that are most important, which can lead to better experimental design in the future.

There are three strategies for building a regression model from a set of possible predictor variables. The simplest is the *forwards* strategy. This picks the best of the predictors and then adds the next best and so on until adding a further variable does not increase the prediction power of the analysis further. The opposite strategy is *backwards* where the analysis starts with all the variables used and then removes the one which has the smallest impact on the predictive power and so on.

The problem with these strategies is that once a variable is added (or subtracted) it remains in (or out) for the rest of the analysis. *Stepwise* regression builds a model in the same way as the *forward* strategy, but each time a variable has been added or subtracted all variable are then considered for addition or removal. In this way, the best fit from any combination of the variables will be made.

The problem for the user of stepwise regression is that decisions have to be made before the test is run. The user must decide how much of an improvement to the prediction is required for a variable to be worth adding to the model and how little loss in the predictive power is allowed for a variable to be dropped.

Stepwise regression is not guaranteed to find the best possible prediction of the 'effect' from the 'cause' variables available. The only way to do this is to try every possible combination of 'causes' and choose the best one. This strategy was once far too time consuming due to the number of calculations required, but, with ever faster computing, it is sometimes offered as an option in statistical packages.

Path analysis

This is a technique related to regression and correlation that removes much of the 'cause' and 'effect' baggage, although it is actually a form of multiple regression. The idea is to generate a map of the interrelationships between variables to visualize the effects and associations within groups of variables. It can be considered as an extension to stepwise regression with multiple possible 'effect' variables. The main difference between path analysis and other techniques with multiple 'causes' is that it allows variables to be correlated, that is, they don't have to be independent. Unfortunately this analysis is not supported in most statistical packages.

9

The tests 3: tests for data exploration

Types of data

Data exploration can be attempted with almost any type of data, although in practice it is most useful where there are a large number of variables and observations. The main aim of data-exploration techniques is to synthesize and process the interrelationships between observations in such a way to make the patterns obvious to the experimenter. Of all the areas of statistics covered in this book this is the one where there is the most scope for personal choice. These techniques are less concerned with *P*-values and should be treated as ways to generate hypotheses rather than test them.

Several of the more commonly employed techniques are considered here. This is certainly not an exhaustive list but it does provide a flavour of the sort of techniques that are available.

Observation, inspection and plotting

The simplest and perhaps most obvious way to generate new hypotheses or to explore relationships between variables is to plot them. Many statistical packages, including SPSS and MINITAB, will produce a matrix of scatterplots with each cell of the matrix having a different plot of two variables. This sort of visual aid will give a general feel for which variables are related to which as well as for the 'shape' of the data. Don't be afraid to experiment with different types of plot before moving on to more standard methods.

Principal component analysis (PCA) and factor analysis

These are two very similar techniques that weight all the available variables to provide the maximum discrimination between individuals. The idea of principal component analysis (PCA; a.k.a. factor analysis, principal axes) is very similar, in many ways, to correlation and regression. The technique can be applied to any

Choosing and Using Statistics: A Biologist's Guide, 3rd Edition. By Calvin Dytham.
Published 2011 by Blackwell Publishing Ltd.

data set that has two or more observations for each individual (e.g. several different morphometric measurements from the same specimen). There are assumptions about the data – that it is continuous and normally distributed – but these can be overlooked if the purpose of the test is to generate further hypotheses.

The technique can be visualized well when there are only two observations for each individual. First imagine a scatterplot of two variables that are correlated such that the points fall within an oval cloud. PCA will determine the line through the points that passes through the long axis of the cloud and will use that as the first principal axis or 'principal component'. A line through the cloud of points at right angles to the first axis will generate a second principal component. Of course this process occurs in multidimensional space within the computer with one dimension for each of the variables included in the analysis and the 'lines' through the clouds of points being formed by weighting each of the variables appropriately.

In this way PCA synthesizes the data from a mass of variables into a set of compound axes. The first axis will explain the most variation, then the second and so on. Therefore inspection of the weightings of the first few axes will show which variables contribute most to the differences between individuals.

In morphometric analysis it is usually the case that individual specimens will vary in size. The first principal axis will nearly always account for size and it is often employed as a method for removing size from the analysis leaving aspects of 'shape' for the second and subsequent axes.

An example

A typical use of PCA and factor analysis is for exploration of morphometric characters. Here there is a very small data set from only 16 individuals of two species of fruit fly: *Drosophila melanogaster* and *Drosophila simulans*. Five morphological characters have been measured to the nearest 0.01 mm, and the sex and species is recorded too. The species are coded 1 and 2 for convenience and sex is coded 1 for female and 2 for male.

It is important to realize that there is no requirement for the grouping variables 'sex' and 'species' as the PCA operates on the measured variables to maximize the differences between individuals (i.e. the rows in this data set). I have used the coded variables only to illustrate the results of the analysis.

There are three main components to the output. The first is the weighting applied to each of the variables to generate the principal axes. The second is the set of eigen values that show how important the principal axes are (each axis will always explain less of the variation than the last). The third is the position of the individuals on the axes: this is used to generate the graphical display of the output (Fig. 9.1).

Investigation of the factor weightings will show which characters are being used to generate the differences between individuals and which are not (the ones with weightings near zero). In the example the first axis (using either PCA

Sex	Species	Thorax length (mm)	Wing length (mm)	Femur length (mm)	Eye width (mm)	Third antennal segment (mm)
1	1	1.01	2.51	0.06	0.52	0.11
1	1	0.98	2.45	0.05	0.53	0.12
1	1	1.02	2.57	0.08	0.55	0.11
1	1	1.05	2.61	0.07	0.52	0.10
2	1	0.98	2.40	0.04	0.54	0.13
2	1	0.89	2.35	0.04	0.50	0.14
2	1	0.89	2.38	0.05	0.50	0.12
2	1	0.95	2.41	0.05	0.49	0.12
1	2	1.20	3.10	0.09	0.48	0.09
1	2	1.15	3.12	0.10	0.52	0.10
1	2	1.18	3.21	0.09	0.52	0.11
1	2	1.21	3.20	0.10	0.55	0.09
2	2	0.95	2.51	0.08	0.56	0.11
2	2	0.94	2.50	0.07	0.49	0.13
2	2	0.96	2.62	0.08	0.51	0.14
2	2	0.91	2.45	0.07	0.52	0.13

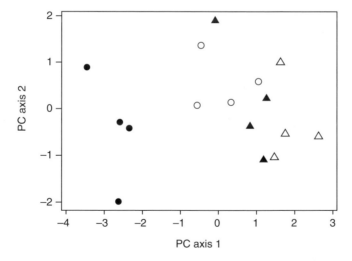

Fig. 9.1 Plotting individuals according to their principal components in MINITAB. PCA maximizes the difference between individuals rather than groups. The individuals are coded by species (shapes) and sex (filled or open shapes). The filled circles are clearly different to the other groups on axis 1 but there is no obvious pattern on axis 2. In many PCA analyses on morphological data axis 1 is closely related to size and other axes to 'shape'.

or factor analysis) is generated by contrasting the three size variables – 'wing', 'thorax' and 'femur' – against 'third antennal segment'. This allows the investigator to focus on these characters as the most important.

The position of individuals on the axes can be used in ANOVA to see if the groups vary. In the example a two-way ANOVA on the first principal component (PC1) using 'sex' and 'species' as grouping variables proved that both factors were highly significant and that they had a significant interaction too. A similar test on PC2 showed no significant differences at all. This shows that there are morphological differences between the sexes and species that are condensed into the first axis.

SPSS Input the data in the same form as the table in the example. Label the columns appropriately. From the 'Analyze' menu, select 'Dimension reduction' then 'Factor…'. In the dialogue box highlight all the measured variables (but not any grouping variables) and move them into the 'Variables:' box. You can click 'OK' now although there are two detours that may pay dividends. Clicking on 'Extraction' will allow you to determine the number of axes ('Factors:') that are generated (I clicked on 'Fixed number of factors:' and selected two 'Factors to extract' for this example). Clicking on 'Scores' allows you to store the scores as variables. If you do this then you can generate a figure similar to the one that I have generated from the example data by using a scatterplot (Fig. 9.1).

The output generated is as follows:

Factor Analysis

Communalities

	Initial	Extraction
thorax	1.000	.932
wing	1.000	.939
femur	1.000	.794
eye	1.000	.994
ant	1.000	.749

Extraction Method: Principal Component Analysis.

Total Variance Explained

Component	Initial Eigenvalues			Extraction Sums of Squared Loadings		
	Total	% of Variance	Cumulative %	Total	% of Variance	Cumulative %
1	3.419	68.386	68.386	3.419	68.386	68.386
2	.988	19.765	88.152	.988	19.765	88.152
3	.383	7.658	95.810			
4	.193	3.857	99.666			
5	.017	.334	100.000			

Extraction Method: Principal Component Analysis.

Component Matrix^a

	Component	
	1	2
thorax	.959	−.112
wing	.952	−.178
femur	.891	.018
eye	.229	.970
ant	−.864	−.046

Extraction Method: Principal Component Analysis.

Of the three output tables, the second and last are useful. In the 'Total Variance Explained' table the summaries for the extracted principal component axes are given. Five 'Components' appear on the left of the table and only two are duplicated on the right, because I chose to have only two factors generated. The maximum number of components, factors or axes will be the same as the number of variables in the data set (five in the example), although that would somewhat defeat the object of PCA. The first 'Component' has the largest eigenvalue and explains the most variance. In the example an eigenvalue of 3.419 translates as nearly 69% of the variation between individuals. The first two 'components' combined explain 88% ('Cumulative %') of the variation between individuals between them.

The last table gives the 'Component Matrix' and is a summary of the weightings assigned to each of the variables for the extracted components or factors. In this case factor 1 ('Component 1') shows a high positive weighting for the first three variables contrasted against a high negative weighting for 'ant' whereas 'component 2' gives a high positive weight to 'eye'. This suggests that the best contrast between individuals is achieved by comparing the 'size' variables with the size of 'ant'. Perhaps individuals with large values for 'thorax', 'wing' and 'femur' tend to have small 'ant' and vice versa.

Sometime weightings are given in unhelpful scientific notation. For example, a weighting of '1.764E-02' translates as 1.764×10^{-2} or 0.01764, a very low weighting.

If the 'save scores as variables' option was used, two new variables will have appeared in the 'Data View' called 'FAC1_1' and 'FAC2_1'. These hold the weighted scores for the first two axes and a scatter of these two variables will allow the generation of figures such as Fig. 9.1: simply scatter these two variables against each other and use the species and sex values to set the symbols.

R There are two functions for PCA in R: 'prcomp()' and 'princomp()'. They have slightly different options but produce similar results. Here I'm assuming that the data have been imported exactly as laid out in the example. It is important that the grouping variables are not considered within the PCA, so I have defined the variables I want to be used in the PCA with the option 'scale=TRUE' which means the effects will sum to one as in SPSS:

```
> summary(prcomp(~thorax+wing+femur+eye+ant,scale=TRUE))
```

```
Importance of components:
                        PC1     PC2     PC3     PC4     PC5
Standard deviation     1.849   0.994  0.6188  0.4391  0.12914
Proportion of Variance 0.684   0.198  0.0766  0.0386  0.00334
Cumulative Proportion  0.684   0.882  0.9581  0.9967  1.00000
```

This output shows how much of the variation is described by each of the principal components as a proportion. In the example 'PC1' explains 68.4%, 'PC2' explains 19.8% and so on.

Next the function 'print()' is used in combination with 'prcomp()' to reveal the weightings given to each of the measurement variables for each principal component:

```
> print(prcomp(~thorax+wing+femur+eye+ant,scale=TRUE))
Standard deviations:
[1] 1.8491400 0.9941184 0.6187849 0.4391321 0.1291438
```

```
Rotation:
            PC1          PC2          PC3          PC4
thorax   0.5185966   -0.11303213   0.1408875   -0.5248445
wing     0.5150173   -0.17934884  -0.2514674   -0.3870185
femur    0.4818698    0.01774549  -0.5946408    0.6000820
eye      0.1236216    0.97599696  -0.0548311   -0.1635158
ant     -0.4672622   -0.04661295  -0.7485390   -0.4334964
```

The first principal component 'PC1', which explains 68% of the variation, weights 'thorax', 'wing' and 'femur' highly positively and 'ant' highly negatively, which indicates that the best separation of individual is achieved by contrasting 'ant' with 'thorax', so animals with a relatively large value of 'ant' and low 'femur' will be at one extreme (note the signs are unimportant). The weightings for PC2 indicate that 'eye' is the most important.

Here the function 'biplot()' is used to produce the output shown in Fig. 9.2:

```
> biplot(prcomp(~thorax+wing+femur+eye+ant,scale=TRUE))
```

MINITAB Input the data in columns as set out in the example. Label all the columns appropriately. From the 'Stat' menu choose 'Multivariate' then 'Principal Components...'. In the dialogue box highlight all the measured variables then click on 'Select'. The analysis will calculate as many components as there are variables (i.e. five in the example) unless you choose a lower number in the 'Number of components to compute:' box. If you wish to keep the position of individuals on each of the axes then you must use the 'Storage'

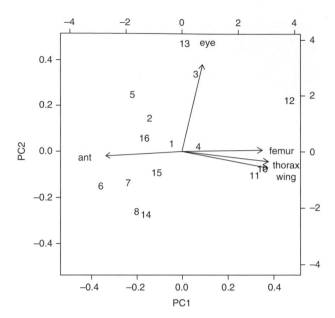

Fig. 9.2 A 'biplot()' in R. Here the locations of each individual on the first two principal components are indicated by numbers (e.g. '10' indicates the location of the tenth individual in the data set). The arrows show the weightings of the variables in the first two principal components.

button then choose columns for values in the 'Scores:' box. In the example I selected two components and then stored them in columns c9 and c10 then scatterplotted c9 and c10 to produce the figure.

If you click OK this is the output you will get:

```
MTB > PCA 'femur' 'thorax' 'wing' 'eye' 'ant';
SUBC> NComponents 2;
SUBC> Scores c9 c10.
```

Principal Component Analysis: femur, thorax, wing, eye, ant

Eigenanalysis of the Correlation Matrix

Eigenvalue	3.4193	0.9883	0.3829	0.1928	0.0167
Proportion	0.684	0.198	0.077	0.039	0.003
Cumulative	0.684	0.882	0.958	0.997	1.000

Variable	PC1	PC2
femur	0.482	-0.018
thorax	0.519	0.113
wing	0.515	0.179
eye	0.124	-0.976
ant	-0.467	0.047

The output confirms the test. Then it gives the eigenvalues of the principal components from one to five. The eigenvalue can be treated as a measure of the variation explained by the axis. Each factor will explain less of the variation than the previous one. This is converted to a proportion of the variation and cumulative proportion for convenience. So the first axis explains 68.4% (0.684) of the variation and the second 19.8%.

In the second part of the output the weightings applied to the five measured variables are given. The first principal component can be interpreted as a contrast between 'ant' and 'thorax', 'wing' and 'femur' while the second is heavily influenced by 'eye'. These weightings maximize the differences between individuals in the sample.

(Or, if the command interface is enabled, type 'PCA' and the list of variables you wish to include at the MTB> prompt in the session window.)

Excel It is not possible to carry out this test in Excel.

Canonical variate analysis

This technique works in very much the same way as PCA but with one crucial difference: the individuals must be assigned to groups before the analysis is run. The test then calculates the variable weightings that will maximize the differences between *groups* rather than individuals, as is the case with PCA.

Canonical variate analysis (or CVA) produces weightings that will allow you to identify those variables that are the most different between groups and discard the ones that are the same. It is important to use real classifications rather than arbitrary ones to get the maximum benefit from the technique.

Discriminant function analysis

As with canonical variate analysis, this technique also requires that the individuals be divided into groups. The idea of discriminant function analysis is to provide a set of weightings that allow the groups to be distinguished. The weightings can then be used on individuals that are not assigned to a group to provide a probability of them belonging to each of the possible groups. If the probability is high then the 'unknown' can be confidently assigned to a group. The power of the weightings is often tested by removing a portion (one or more observation) from the data set, using the remainder to create the weightings and then to use the weightings to assign the removed individuals to groups. The hit rate is a measure of the power of the test to discriminate real unknowns.

An example

We will use the same example as for the PCA but with one of the grouping variables as the target discriminator. The analysis will determine how often an individual of unknown species or sex would be attributed to the correct group.

SPSS Input the data in columns and label appropriately. From the 'Analyze' menu select 'Classify' then 'Discriminant…'. Move the variable with the group codes into the 'Grouping variable:' box. Click on 'Define range…' to input the lowest and highest group code numbers. (In this case I used the variable 'sex' and defined groups 1 and 2). Click 'Continue' to return to the 'Discriminant Analysis'. Next move all the measured variables into the 'Independents:' box.

It is often important to visit the 'Classify' dialogue before the test is run. There the 'Prior probabilities' should be changed to 'Compute from group sizes' unless you are sure that unknown individuals should have an equal chance of being in either group (in the example I left 'All groups equal' as I was classifying to sex.). Also on the 'Classification' dialogue the very useful option of 'Leave-one-out classification'. Output generated with this option selected is shown below.

There is now a range of possible options that can be selected from the buttons at the bottom. The most useful is 'Save…' where the 'Predicted group membership', 'Probabilities of group membership' and 'Discriminant scores' can be stored as separate variables. If you don't do this then there is no way of checking which individuals have been assigned to each group which is what determines the accuracy rate of the discriminant analysis. Click 'OK' and masses of output appears in the 'Output' window of which this is an edited section.

Standardized Canonical Discriminant Function Coefficients

	Function
	1
thorax	2.076
wing	−1.659
femur	.170
eye	.090
ant	−.261

Structure Matrix

	Function
	1
thorax	.770
ant	−.744
wing	.523
femur	.370
eye	.139

Functions at Group Centroids

	Function
sex	1
1	1.524
2	−1.524

The tables shown are first 'standardized coefficients' where the weightings given to each of the measured variables to maximize the differences between groups are given. Clearly 'thorax' and 'wing' are the highest weighted in the example. The 'Structure Matrix' gives a slightly different view of the data as it appears that although 'thorax' has the highest correlation with the discriminating function the negative correlation for 'ant' is almost as important once the relatively small size of the measurements has been accounted for.

Finally comes the 'group centroids' for the two groups on the axis of the discriminant function. As there are only two groups and they are of equal size in the example it is not surprising that the group centroids (mean position) are the same distance either side of zero. In practice this means that any individual that has the weightings applied to the measurement variables and scores more than zero will be assigned to group 1 and less than zero to group 2.

With the 'Leave-one-out classification' selected the following output also appears:

Classification Statistics

Prior Probabilities for Groups

sex	Prior	Cases Used in Analysis	
		Unweighted	Weighted
1	.500	8	8.000
2	.500	8	8.000
Total	1.000	16	16.000

Classification Results[b,c]

			Predicted Group Membership		
		sex	1	2	Total
Original	Count	1	8	0	8
		2	1	7	8
	%	1	100.0	.0	100.0
		2	12.5	87.5	100.0
Cross-validated[a]	Count	1	7	1	8
		2	2	6	8
	%	1	87.5	12.5	100.0
		2	25.0	75.0	100.0

a. Cross validation is done only for those cases in the analysis. In cross validation, each case is classified by the functions derived from all cases other than that case.

b. 93.8% of original grouped cases correctly classified.

c. 81.3% of cross-validated grouped cases correctly classified.

Here the 'Prior Probabilities' confirms that there was a 0.5 probability of being assigned to either group (i.e. 50% female, 50% male). The analysis gives the results of running the analysis once for each individual with the target individual excluded and then reassigned to a group based on the discriminant function score. The 'Classification Results' table shows that all of group 1 (females) were assigned to group 1, but only seven of eight in group 2 were assigned to the correct group. This gives an overall efficiency of 93.75% (here reported as 93.8%) for the discrimination of groups.

The group assignments for each individual are made in a new variable ('dis_1') in the data window. In the example the one misclassified individual, row 5, should have been in group 2 but was assigned to group 1. It has a 'dis1_1' (discriminant function score) of 0.032 and had a 52% chance of being in group 1 and 48% of group 2. All other individuals were assigned to the correct group.

R To carry out a discriminant function analysis requires an extra library or package to be installed in R. There are many, many options available. Here the library 'MASS' has been used to carry out an analysis using the function 'lda()' the output of which has been assigned to 'fit':

```
> library(MASS)
> fit<-lda(species~thorax+wing+femur+eye+ant,CV=TRUE)
> fit
$class
[1] mel mel sim mel mel mel mel mel mel sim sim sim mel
sim sim sim

Levels: mel sim

$posterior
        mel              sim
1       9.992298e-01     7.701998e-04
2       9.999105e-01     8.953673e-05
...
```

There is a lot more output. The most interesting sections are the '$class' where the identifications for each individual in the data set are given. The '$posterior' section gives the probability of membership to each group. In this example the first individual is 99.92% likely to be 'mel' and 0.077% likely to be 'sim'.

You can make a table of the number of fits to each species and then report using 'diag()' to make an object I have called 'ct'. This can be used to display what proportion of each species are correctly identified using 'prop.table()', followed by the proportion of correct identifications overall:

```
> ct<-table(fly$species,fit$class)
> diag(prop.table(ct,1))
mel          sim
0.875        0.750
> sum(diag(prop.table(ct)))
[1] 0.8125
```

Here 87.5% of 'mel' were correctly assigned and 81.25% overall.

MINITAB Input the data into columns and label appropriately. From the 'Stat' menu select 'Multivariate' then 'Discriminant analysis…'. Move the grouping variable ('sex' or 'species' in the example) into the 'Groups:' box. Highlight all the measured variables and click 'Select' to move them into the 'Predictors:' box. Click 'OK'. (In the example I chose 'sex' as the grouping variable as the analysis was 100% accurate when 'species' was used.)

You will get the following output:

Discriminant Analysis: sex versus wing, femur, thorax, eye, ant

```
Linear Method for Response: sex
Predictors: wing, femur, thorax, eye, ant

Group      f      m
Count      8      8

Summary of classification

                   True        Group
Put into Group      f            m
f                   8            1
m                   0            7
Total N             8            8
N correct           8            7
Proportion        1.000       0.875

N = 16   N Correct = 15   Proportion Correct = 0.938

Squared Distance Between Groups
                   f            m
f              0.00000      9.28891
m              9.28891      0.00000

Linear Discriminant Function for Groups
                   f            m
Constant        -795.5       -759.6
wing            -117.7        -97.2
femur           -264.2       -293.4
thorax           927.6        838.2
eye             1155.6       1143.9
ant             3098.6       3174.4

Summary of Misclassified Observations

Observation   True      Pred    Group    Squared
              Group     Group            Distance    Probability
5**             m         f         f      9.183        0.524
                                    m      9.378        0.476
```

The output confirms the test and then the variables used for grouping and prediction. It then gives the number of observations in each of the groups (in the example, there were eight females and eight males). Next comes a 'Summary of Classification' where the real group 'True group' and the group that each individual is put into is compared. In the example all of group 'f' were correctly assigned to group 'f' but one individual from group 'm' was incorrectly assigned to group 'f'. A summary of the accuracy comes next giving it as a proportion correct of 0.938 in the example (i.e. 93.8% or 15/16 of the individuals were assigned to the correct group). The squared difference between the groups is not very useful for just two groups but will show which groups are most similar to which if there are more than two. Then comes a section labelled 'Linear Discriminant Function for Groups' which gives weightings to each of the measured variables that are akin to the slopes in a multiple regression. Finally is a summary of the misclassifications. In the example only one of the 16 was misclassified. The individual was in row 5. The analysis gave it a 52.4% probability of being in group 'f' (i.e. wrongly assigned) and 47.6% of group 'm' (i.e. correctly assigned) so it was a borderline case.

The analysis suggests that an unknown individual from this area could be assigned to 'sex' from analysis of the measured variables alone with an accuracy of over 93%.

Excel You cannot carry out this test in Excel.

Multivariate analysis of variance (MANOVA)

All the ANOVA techniques that are introduced in Chapter 7 have only one observed variable although they had one or more classification variables. Therefore they can all be called *univariate* analyses. The statistical analysis MANOVA allows more than one observed variable to be analysed at once; hence it is a multivariate test. This test often has the effect of combining two or more rather borderline significant results into a highly significant single result.

An example

Using the data for fly morphology again (see above) a MANOVA is used on two response variables – 'thorax length' and 'wing length' – simultaneously with 'species' being used as the factor. In two univariate ANOVAs on these data 'thorax' is not significant, while 'wing' is significant at $P=0.01$. The MANOVA uses information from both variables to give a highly significant result of $P=0.001$.

SPSS From the 'Analyze' menu, select 'General Linear Model' then 'Multivariate'. Move the observation data to the 'Dependent Variables:' box and the main effect ('Species' in the example) to the 'Fixed Factor(s):' box. Click 'OK'.

Multivariate Tests[b]

Effect		Value	F	Hypothesis df	Error df	Sig.
Intercept	Pillai's Trace	.992	817.233[a]	2.000	13.000	.000
	Wilks' Lambda	.008	817.233[a]	2.000	13.000	.000
	Hotelling's Trace	125.728	817.233[a]	2.000	13.000	.000
	Roy's Largest Root	125.728	817.233[a]	2.000	13.000	.000
species	Pillai's Trace	.674	13.447[a]	2.000	13.000	.001
	Wilks' Lambda	.326	13.447[a]	2.000	13.000	.001
	Hotelling's Trace	2.069	13.447[a]	2.000	13.000	.001
	Roy's Largest Root	2.069	13.447[a]	2.000	13.000	.001

a. Exact statistic
b. Design: Intercept + species

Tests of Between-Subjects Effects

Source	Dependent Variable	Type III Sum of Squares	df	Mean Square	F	Sig.
Corrected Model	thorax	.033[a]	1	.033	3.163	.097
	wing	.574[b]	1	.574	8.933	.010
Intercept	thorax	16.545	1	16.545	1570.996	.000
	wing	112.307	1	112.307	1748.382	.000
SPECIES	thorax	.033	1	.033	3.163	.097
	wing	.574	1	.574	8.933	.010
Error	thorax	.147	14	.011		
	wing	.899	14	.064		
Total	thorax	16.725	16			
	wing	113.780	16			
Corrected Total	thorax	.181	15			
	wing	1.473	15			

a. R Squared = .184 (Adjusted R Squared = .126)
b. R Squared = .390 (Adjusted R Squared = .346)

There are three tables of output. In the first (not shown) the number of observations in each group is confirmed. In the second output comes the results of the MANOVA labelled 'Multivariate Tests'. The section on the 'intercept' is not very informative; it is the bottom half with the effect labelled as 'species' where the results come. There are four different MANOVA statistics (Pillai's Trace etc.) and, in this case, they all give the same results. For all four the *P*-value (labelled 'Sig.') is 0.001 and the *F*-ratios and degrees of freedom are identical too.

In the final table the univariate ANOVAs on the same data are shown. In this case the important lines are with the source labelled 'species'. One of the ANOVAs has a *P*-value of 0.097 (i.e. not significant) while the other has *P*=0.010 which is significant, but not as conclusive as the result from the MANOVA.

R I am assuming that the data are arranged as for the PCA. This method uses very similar syntax to ANOVA, but first combines the response variables. In the example the two variables have been combined into an object called 'Y':

```
> Y <- cbind(thorax,wing)
```

This is then used in an ANOVA model; here the simplest version is used with a single variable as a factor. The output from the 'manova()' function is put into an object 'fit':

```
> fit <- manova(Y ~ species)
```

The output is then visualized using 'summary()'. This is done twice, first giving a separate ANOVA table for each response variable. The second gives the MANOVA version of the output and asks for the 'Wilks' version of the test (other options, 'Roy' and 'Pillai', will give similar results):

```
> summary.aov(fit)
Response thorax:
            Df  Sum Sq     Mean Sq    F value   Pr(>F)
Species     1   0.033306   0.033306   3.1626    0.09706
Residuals   14  0.147438   0.010531
---
Signif. codes:    0 '***' 0.001 '**' 0.01 '*' 0.05 '.'
0.1 ' ' 1

Response wing:
            Df  Sum Sq     Mean Sq    F value   Pr(>F)
species     1   0.57381    0.57381    8.9329    0.009766 **
Residuals   14  0.89929    0.06423
---
Signif. codes: 0 '***' 0.001 '**' 0.01 '*' 0.05 '.' 0.1 ' ' 1
> summary(fit, test="Wilks")
            Df Wilks approx F  num Df  den Df  Pr(>F)
species     1   0.32586  13.447    2    13    0.0006834 ***
Residuals   14
—
Signif. codes: 0 '***' 0.001 '**' 0.01 '*' 0.05 '.' 0.1
' ' 1
```

The univariate outputs confirm that 'thorax' is not significant with species as a factor ($P=0.097$) and that 'wing' is significant ($P<0.01$). The second output gives the result when the two variables are combined in the MANOVA. Here the significance is greater, with $P<0.001$.

MINITAB From the 'Stat' menu, select 'ANOVA' then 'General MANOVA…'. Move two or more variables to the 'Responses:' box. Move the main effect ('species' in the example) to the 'Model:' box. Click 'OK'.

The following output appears:

General Linear Model: thorax, wing versus species

```
MANOVA for species              s = 1   m = 0.0        n = 5.5

                                                    DF
Criterion           Test Statistic      F    Num   Denom        P
Wilks'                     0.32586  13.447    2      13    0.001
Lawley-Hotelling           2.06883  13.447    2      13    0.001
Pillai's                   0.67414  13.447    2      13    0.001
Roy's                      2.06883
```

This, rather meagre, output can be augmented by requesting various options. A useful option is to run the univariate ANOVAs on all the variables used in the MANOVA to see how much better the groups are separated. In the output there are three different methods of calculating the significance of a MANOVA. As is clear from the output they all provide the same conclusion with identical 'P', 'F' and 'DF' values. In each case the P-value is 0.001, indicating that the two species are highly significantly different.

Excel There is no method for carrying out this test in Excel.

Multivariate analysis of covariance (MANCOVA)

This technique is related to MANOVA in the same way that ANCOVA is related to ANOVA. If there are more than one observed variables, one or more ways of classifying the data and furthermore there is a measured observation that is known to have an effect on the observations then MANCOVA can be used to remove the effect of this confounding variable from the analysis.

To carry out this test in SPSS or MINITAB, follow the MANOVA example and add a covariate to the box indicated. In R you can add a covariate to the function as in the ANCOVA example.

Cluster analysis

This is a general term for a huge range of techniques for the classification of individuals. These techniques are becoming increasingly important as more detailed statistical analysis of DNA sequence analysis is being attempted. Cluster analysis can be used to generate dendrograms that show putative phylogenetic relationships or at least divide individuals into groups that might have taxonomic meaning. However, cluster analysis is certainly not restricted to molecular sequence analysis; it has a long history in taxonomy (both cladistics and phenetics) and in community ecology (particularly in vegetation classification or ordination) and is now becoming a key technique in the study of gene expression and gene families.

In its simplest form cluster analysis can be imagined as a step-by-step process. First the individuals are depicted as a scatter of points. Then the individuals that are closest together are identified and their similarity recorded as the distance between them. The two points are amalgamated into a single point located half way between the two. The next two closest points are then identified and amalgamated. This process can be continued until there is only one point.

SPSS There are several methods available. As for discriminant function analysis and PCA, input the data in columns so that each individual is represented by a row of observations. From the 'Analyze' menu select 'Classify' and then choose 'Hierarchical Cluster'. Simple dendrograms can be generated by selecting the sub-menu 'plots' and checking the 'Dendrogram' option. I suggest that for 'Icicle' you select 'None' as this display is very difficult to interpret.

Using the example data and the 'Method' of 'Within-groups linkage' and 'Rescale to 0-1 range' option I generated the following agglomeration schedule:

Average Linkage (Within Groups)

Agglomeration Schedule

	Cluster Combined			Stage Cluster First Appears		
Stage	Cluster 1	Cluster 2	Coefficients	Cluster 1	Cluster 2	Next Stage
1	6	7	.000	0	0	13
2	11	12	.001	0	0	12
3	2	5	.002	0	0	4
4	2	8	.003	3	0	9
5	3	4	.003	0	0	10
6	14	16	.003	0	0	9
7	9	10	.004	0	0	12
8	1	13	.005	0	0	11
9	2	14	.006	4	6	11
10	3	15	.007	5	0	14
11	1	2	.008	8	9	13
12	9	11	.010	7	2	15
13	1	6	.013	11	1	14
14	1	3	.026	13	10	15
15	1	9	.266	14	12	0

This shows that the first pair of individuals to be clustered (i.e. the most similar pair) were individuals 6 and 7. These will then form a new group, given the number of the lowest representative in the group, that is compared to the remaining individuals. The next most similar pair were individuals 11 and 12, then 2 and 5 and so on. This process continues until in step 15 a group of 12 individuals including individual 1 is linked to a group comprising the remaining four individuals including individual 9.

The crude dendrogram shows this graphically, but as it uses ASCII symbols the resolution is very poor and it is certainly not a chart that can be used in a report.

R There are many clustering methods available in R. Here is a very simple one that produces output similar to MINITAB. First the variables that are going to be used are bound into a single object here called 'Y'. If you are going to use all the data in a dataframe then you can skip this bit:

```
> Y <- cbind(thorax,wing,eye,femur,ant)
```

A hierarchical clustering using 'Ward's method' is carried out by first calculating the Euclidean distances between individuals in the 'dist()' function and putting the results into an object 'd', then performing the clustering using 'hclust()' on 'd' and putting the results into 'fit'. The output is visualized using 'plot()':

```
> d<-dist(Y, method="euclidean")
> fit<-hclust(d,method="ward")
> plot(fit)
```

This produces output very like the MINITAB output in Fig. 9.3 (see below). To add a flourish, here is some code that will add four red rectangles to the plot. Change 'k=4' if you want a different number of groups.

```
> groups<-cutree(fit, k=4)
> rect.hclust(fit, k=4, border="red")
```

In the example this puts a box around the group 9, 10, 11 and 12, but fails to identify the other three groups accurately. However, this is a very small data set with rather few morphometric characters measured and this method of clustering is a very simple one.

MINITAB There is a variety of clustering methods available in MINITAB and graphical output of reasonable quality can be generated. Input the data in columns so that each individual is represented by a single row. From the 'Stat' menu select 'Multivariate' and then one of the three 'Cluster' methods. To generate the analysis that produced the dendrogram shown in Fig. 9.3, I used the 'Cluster observations…' method then moved the measured variables from the example data for PCA to the 'Variables or distance matrix:' box.

(Or, if the command interface is enabled, and assuming data are in columns c1 to c7, type 'CLUO c3 c4 c5 c6 c7' at the MTB> prompt in the session window.)

Cluster Analysis of Observations: thorax, wing, femur, eye, ant

```
Euclidean Distance, Single Linkage
Amalgamation Steps
```

Step	Number of clusters	Similarity level	Distance level	Clusters joined		New cluster	Number of obs. in new cluster
1	15	95.9016	0.037417	6	7	6	2
2	14	94.6340	0.048990	11	12	11	2
3	13	94.2040	0.052915	2	5	2	2
4	12	93.4280	0.060000	3	4	3	2
5	11	93.3373	0.060828	2	8	2	3
6	10	92.8174	0.065574	14	16	14	2
7	9	92.5711	0.067823	2	14	2	5
8	8	92.5711	0.067823	2	6	2	7
9	7	92.4908	0.068557	9	10	9	2
10	6	92.4113	0.069282	1	2	1	8
11	5	92.2548	0.070711	1	3	1	10
12	4	91.8033	0.074833	1	13	1	11
13	3	89.8423	0.092736	1	15	1	12
14	2	89.4939	0.095917	9	11	9	4
15	1	43.6461	0.514490	1	9	1	16

```
Final Partition
Number of clusters: 1
```

	Number of observations	Within cluster sum of squares	Average distance from centroid	Maximum distance from centroid
Cluster1	16	1.67181	0.274649	0.585678

The output shows how the clusters are constructed and how similar individuals are. In the example the two most similar individuals are 6 and 7; they are joined to form a cluster which is given the number of the lowest numbered individual in the cluster. This cluster is later joined to a group containing individual 2 at step 8. Eventually all individuals are joined into a single group; in the example the final step is to join a group of four individuals (9, 10, 11 and 12) to the rest. This is displayed in the dendrogram in Fig. 9.3.

Excel No clustering techniques are available in Excel.

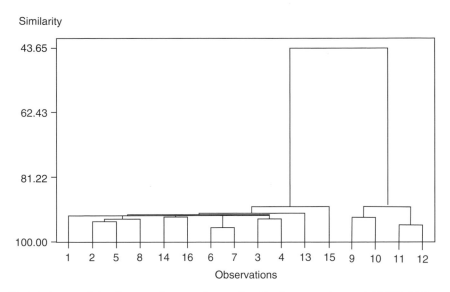

Fig. 9.3 A typical dendrogram showing the results of a cluster analysis in MINITAB. This figure was generated using the 16 individuals used as the example data for PCA. There are some clear groupings. For example, individuals 9–12 form a separate group on the right of the diagram.

DECORANA and TWINSPAN

DECORANA (detrended correspondence analysis) and TWINSPAN (two-way indicator species analysis) are two analyses developed at the Institute for Terrestrial Ecology (now the part of the Centre for Ecology and Hydrology, UK) that are now widely used in the comparison of communities from species abundance data and in the exploration of ecological data.

TWINSPAN, as suggested by its name, is very useful for comparing list of species from sets of sites. For example, in a study of woodland carabids the presence and absence data, or measures of abundance, generate a discriminating function that can group similar woodlands and generate dendrograms. Both are available in R.

Symbols and letters used in statistics

One of the surest ways of making a statistics book difficult to read is the tendency to use Greek letters, single italicized letters or obscure symbols. As far as possible I have tried to avoid these things in this book. Here are the ones that you are most likely to encounter.

Greek letters

These are often used to signify the true values of particular statistics (i.e. the value you would get if you were able to measure the entire population rather than a sample). The estimates you get of the true values are often then labelled with the corresponding normal letter.

Π (pi) product of the terms following it (multiply together)
π (pi) a constant (3.142) used in geometry
Σ (sigma) sum of the terms following it (add up)
α (alpha) the critical significance level for the rejection of a hypothesis (usually 0.05)
β (beta) true regression coefficient (estimated by the statistic, b)
χ (chi) χ^2 is a commonly encountered statistical distribution
γ (gamma) γ_1 is the true value of skewness; γ_2 is the true value of kurtosis
μ (mu) true mean of a population
ρ (rho) true correlation coefficient (estimated by the statistic, r)
σ (sigma) the true standard deviation of a population
σ^2 (sigma squared) the true variance of a population
T (tau) the statistic of Kendall rank-order correlation
Δ or δ (delta) increment (tiny difference or change)

Symbols

$-$ (overbar) indicates a mean
$\sqrt{}$ square root
$=$ is equal to

Choosing and Using Statistics: A Biologist's Guide, 3rd Edition. By Calvin Dytham.
Published 2011 by Blackwell Publishing Ltd.

$\equiv$	is identically equal to
$\neq$	is not equal to
$\approx$	is approximately equal to
$\sim$	is distributed as
$\sim$	used in R to separate predictor from response in a statistical model
$\|\ \|$	absolute value of the number between the bars; e.g. $\|-6\| = 6$
!	factorial (e.g. $3! = 1 \times 2 \times 3 = 6$)
[ ]	used in R for matrix notation
( )	used in R to enclose arguments sent to a function
$\pm$	plus or minus (so 5 ± 2 means a range from 3 to 7)
$\geq$	is greater than or equal to
$<$	is less than (points to smaller value)
$\ll$	is much less than
$>$	is greater than (points to smaller value)
$\gg$	is much greater than
<-	used in R to assign the output from a function
^	used in some statistical packages (e.g. Excel, R) to mean 'raise to the power of'
$\cap$; $\cup$; $\subset$; $\not\subset$	symbols used in set work (intersection; union; is a subset of; is not a subset of)
:	used in R to indicate an interaction term
.	used in R to indicate a nearly significant result, $P > 0.05$ but $P < 0.1$
*	indicating a significant result (usually a P-value is flagged at <0.05)
*	used in statistical packages to mean 'multiplied by'
*	used in R between factors to indicate that main effects and interactions are required
**	denotes a highly significant result (usually $P < 0.01$)
**	used in some statistical packages (e.g. SPSS) to mean 'raise to the power of'
***	denotes a very highly significant result (usually $P < 0.001$)
_____	used to underline groups that are not significantly different (see *Post hoc* tests, page 138)
∞	infinity (an infinite number), often used in statistical tables to indicate the value of an asymptote
$\propto$	varies directly as, is proportional to
$\Rightarrow$	implies
$\times$ or	multiply
$\therefore$	therefore

Upper-case letters

CI confidence interval
CL confidence limit

CV coefficient of variation
d.f. degrees of freedom (also DF or df)
F F-value (e.g. the output from ANOVA), the ratio of within- and between-group variance
F sometimes used to indicate a function
H_0 null hypothesis (the uninteresting hypothesis: nothing is happening)
H_1 alternative hypothesis (the interesting hypothesis: something is happening)
MS mean square (SS/df in an ANOVA table)
P probability (more usually P, p or p)
PI prediction interval
SD standard deviation (also s)
SE standard error
SS sum of squares
X^2 estimate of value for χ^2 (chi-square)

Lower-case letters

a the intercept of a regression line (where the line crosses the y-axis)
a.k.a. also known as; not statistics used several times in this book
b slope of a regression line
d.f. degrees of freedom (sometimes df or DF)
e a constant ($=2.172$) used as the base for natural or Naperian logarithms (ln)
g estimate of value of γ (gamma); $g_1 = $ skew, $g_2 = $ kurtosis
f used to indicate a function
i often used to indicate a sequence of observations (e.g. x_i)
j often used to indicate a second sequence of observations (e.g. x_{ij})
m often used to indicate the sample mean
p probability (also P, P or p)
p binomial probability (e.g. 0.5 probability of an individual being female)
r measure of correlation (Pearson product-moment correlation, varies from -1 to 1)
r_s measure of correlation produced by Spearman's rank-order correlation
r^2 a measure of the amount of variation accounted for by a regression line or correlation
s standard deviation of a sample (also SD)
s^2 variance of a sample
t value of the statistic resulting from a Student's t-test
v occasionally used to indicate variance of a sample
x often used to indicate an observation
y often used to indicate a second observation on the same individual as x
z often used to indicate a third observation on the same individual as x and y

Glossary

An explanation of commonly encountered words, concepts and acronyms including thumbnail description of many statistical tests.

a posteriori A phrase applied when a hypothesis is generated after the data have been collected. Sometimes used as a synonym for a *post hoc* test.

a priori A phrase applied when a hypothesis is generated before the data have been collected.

abscissa The x or horizontal axis of a graph.

accuracy The closeness of a measure to its true value (different to precision).

alpha The name usually given to the critical value of P required to reject a null hypothesis; i.e. usually 0.05.

ANCOVA The short term for analysis of covariance.

Anderson–Darling test A statistical test used to determine whether a set of data are normally distributed.

angular transformation A synonym of arcsine transform. A transformation that is often used to 'normalize' percentage or proportion data.

ANOVA The short term for analysis of variance; originally coined by Tukey.

ANOVAR The alternative term for analysis of variance.

arcsine square-root transformation *see* arcsine transform

arcsine transform An operation that is often used to 'normalize' percentage data; a synonym for arcsine square-root transformation and angular transformation. *Note*: arcsine is sometimes written as arcsin.

arithmetic mean A measure of position: the sum of all values divided by the number of observations (synonym of mean).

association The measure of the strength of the relationship between two variables. Often used as a synonym of correlation.

assumptions Many statistical tests assume that the underlying distribution of the measurements of a whole population is of a particular type or that measurements are made without error, etc.

asymptote The value of a line that a curve is approaching but never meets.

attribute When used to describe an observation it usually implies a small number of possible categories that have no meaningful sequence; for example, 'blue', 'pink' and 'white'.

Choosing and Using Statistics: A Biologist's Guide, 3rd Edition. By Calvin Dytham.
Published 2011 by Blackwell Publishing Ltd.

autocorrelation A common problem with sets of observations is that they are not truly random and are more similar to their 'neighbours' in space or time than is expected by chance.

average A synonym for arithmetic mean (used by Excel for arithmetic mean).

back transformation A process by which observations are transformed back to their original units.

balanced design An experiment where the same number of observations are made for each factor level (or factor combination in two-way ANOVA).

bar chart A graph to display the distribution of a discrete variable where each category on the x-axis represents one possible value.

Bartlett's test A test for homogeneity of variance.

Bartlett's three-group method A model II regression technique.

Bayesian statistics A whole branch of statistics based on a different approach to the concepts of probability and certainty.

Bernoulli distribution Values can only be one of two possibilities (e.g. 0 or 1).

best fit The statistical model which explains the most variation. The best fit is often constrained by the type of statistical test. For example, if the model is a linear regression then the 'best fit' will be the straight line which accounts for the most variation (leaves the smallest residual variation).

between-sample variance A synonym for between-groups sum of squares. In ANOVA and regression it is a measure of the variation between factor levels. Comparison of the between-sample variance and within-sample variance is how ANOVA works.

bias Any systematic error in measurement.

bimodal A frequency distribution that has two peaks.

binomial distribution A theoretical probability distribution of events that can occur in two categories. Can be used as a null hypothesis to determine whether given data are random or not.

bivariate Tests which are applied to two variables.

bivariate normal An assumption of many bivariate parametric tests is that the two variables are both normally distributed.

Bonferroni method A *post hoc* test, used after a one-way ANOVA, to determine which groups are different from which. More generally a method for reducing the critical value of alpha (normally 0.05) when many tests are carried out in the same experiment: divide 0.05 by the number of tests to get new value of alpha.

bootstrapping A method of analysis that improves many estimated statistics by using repeated subsamples of a data set.

box plot A synonym of box and whisker.

box and whisker A method of displaying data where a horizontal line represents the median, a box extends to cover the interquartile range and a line may extend away from the box to the extreme values.

calibration curve Regression can predict y from any value of x. This can be used to generate a calibration curve.

canonical variate analysis A multivariate test where the data are divided into groups and the test weights variables to maximize differences between the groups.

caption A small piece of text that should accompany *every* figure and table used in a report, making it interpretable without recourse to the main text (synonym for legend).

categorical A description of data or a variable that is only described by categories. This can be as simple as 'red, blue, green'. Categories are often called attributes.

causation The assumption in statistical tests that the value of the independent variable causes the result in the dependent variable.

chi-square (χ^2) distribution A distribution widely used in statistics that is based on the deviation of sample variance from the true variance of a population. A χ^2 distribution can be generated from a population of standard normal deviates as the probability density of a very large number of samples scaled thus: $((n-1)s^2)/\sigma^2$.

chi-square goodness of fit A chi-square test where the expected values are taken from a particular distribution against which the observed distribution is being compared.

chi-square test A contingency-table-based statistical test to explore hypotheses of association between variables; expected values are generated by the table.

cluster analysis A number of multivariate tests that group observations by similarity and may provide insight into the data.

component A synonym for a factor or axis in a principal components analysis.

confidence interval A measure of spread: it uses the standard error and the t-distribution to give a range of values within which there is a percentage probability of the true mean occurring (usually using set at 95%). In regression the 95% confidence intervals of slope and intercept are used to generate 95% confidence intervals of the line. This zone is not parallel with the best fit line but wider at extreme values of x.

confounded design This often occurs in a multifactorial design when not all combinations have been used or when the effects of one factor cannot be disentangled from the effects of another (i.e. they are not independent). Can be due to lack of space or resources a poor experimental design.

conservative test A test where the chance of type I error is reduced and that of a type II error increased (i.e. the null hypothesis is rejected less often than it should be).

constant Any fixed value. In regression analysis the intercept (value of y when x is zero) is sometimes called this.

contingency table A way of displaying data that has been assigned to categories for two variables; for example, broadleaf/conifer and insect damage $> 10\%$? yes/no.

continuity correction A method of correcting bias in various contingency tests. Makes test results more conservative. Also called Yates' correction.

continuous A description of data or variables indicating every value is possible (in theory); for example, any linear measurement; >29 possible values as a rule of thumb.

control A general term given to the section of an experiment which is unmodified by the experimenter.

Cook's distance statistic A useful measure of the influence of single measures on the outcome of regression analysis (any value > 1 is large).

correlation A method of measuring the association between two variables (often used as a synonym for Pearson product-moment correlation). Warning: large samples will nearly always produce mathematically 'significant' results even when the biological is only miniscule.

covariance Technically, this is the sum of the product of the deviations from the mean for a set of paired observations (i.e. the variance that remains once the relationship between two variables is accounted for). Related to variance and useful in many statistics.

Cox regression A type of regression used for assessing hazards (e.g. mortalities) to make predictions about the chance of events occurring.

Cramér coefficient The result of a test on contingency tables other than 2×2.

Cramér's V A statistic correcting chi-square values for sample size.

cumulative frequency The number of times an observation takes a particular value *or less* in a data set; for example, if there were 26 zeroes, 22 ones and 15 twos then the cumulative frequency for two would be $26 + 22 + 15 = 63$.

cumulative probability The probability of achieving a particular value and all values greater (or smaller). A type of data display where probabilities are accumulated from left to right until all observed values have been included.

data Observations recorded during an experiment or survey (plural).

datum A single observation.

degrees of freedom A number related to the sample size that accounts for the number of observations made, number of factor levels and any manipulations carried out. Degrees of freedom are lost, for example, when a set of values is constrained to total 1 or when a set of data is being compared to a standard distribution with mean or other parameters the same as the data.

dependent variable The observed or 'effect' variable not set by the experimenter.

derived A value that results from a combination of two or more values; for example, ratio, proportion or percentage.

descriptive statistic Anything that summarizes the data; for example, mean or standard deviation.

dichotomous variable A variable that can only take two possible values (e.g. yes/no).

discontinuous data Only a limited number of values are possible (<30 as a rule of thumb).

discontinuous variable A variable that only has a limited number of values (usually but not always integers).

discrete A description of data or variables indicating that not every value is possible; for example, the number of children can never be 2.5.

discriminant function analysis A multivariate test that assigns individuals to groups.

dispersion The way in which the data are distributed, often measured by standard deviation; a synonym of spread.

distribution The spread of a set of observations.

double-blind An experiment when the experimenter does not know which treatment is being applied until labels are decoded. A very good way to avoid inadvertent bias by recorders and a requirement of all experiments on humans.

dummy data Made up before the experiment is carried out used for a dry run of the statistical tests planned.

Duncan's method A *post hoc* test, used after a one-way ANOVA to determine which groups are different from which.

Dunn–Sidák method A *post hoc* test, used after a one-way ANOVA to determine which groups are different from which. More generally it is a method of setting the critical value of alpha (usually set to 0.05) when many experiments or tests are carried out at the same time. This method is often considered superior to the Bonferroni method (q.v.): new alpha equals $0.95^{1/k}$, where k is the number of tests that have been carried out.

Dunn test One of many *post hoc* tests used in ANOVA to determine which groups are different from which.

error A word quite widely used in statistics. Can be applied to the degree of accuracy with which measurements are made. Can also be used in statistical tests such as ANOVA to refer to variation not accounted for by the model being tested.

estimate Any statistics (e.g. mean) calculated from a sample is an estimate of that value for the whole population.

expected frequency The occurrence of an event determined from a theoretical distribution or hypothesis based on previous observations.

expected value Values needed in contingency tables when they are calculated using the row and column totals or values derived from a theoretical distribution or hypothesis.

experiment Random assignment of subjects to controlled 'experimental' conditions.

experimental design A process of planning which should always occur before an experiment begins to maximize the usefulness of results obtained while minimizing the effort.

exponential distribution A distribution that arises when there is a constant probability of an event occurring; for example, birth rate or radioactive decay rate. Logarithmic transformations will turn exponential curves into straight lines.

extrapolation Whenever predictions are made beyond the range of the data available

F-distribution An asymmetric, continuous distribution, used in ANOVA tests with a modal value of 1, representing the frequency of occurrence of the ratio of between-group variance/within-group variance for groups with the same distribution. Called F in honour of R.A. Fisher, the originator of ANOVA.

F-ratio A synonym for F-value.

F-test A statistical test with the null hypothesis that groups have the same variance. Often offered as a test for homogeneity of variance in a t-test.

F-value The output from an ANOVA test. The ratio of between-group variance and within-group variance.

factor A grouping variable or independent variable. For ANOVA factors must be discontinuous/categorical or made to appear so.

factor level A number representing different groups; for example, sap flows are examined from three species of tree. The species being groups that are arbitrarily assigned numbers as factor levels.

factorial (experiment) An experiment or survey where there are two or more factors and each possible combination of factor levels is represented in the data.

field experiment A general term given to any experiment where conditions are manipulated by the experimenter which occurs outside a controlled laboratory.

figure Any graphical item in a report.

first-order interaction An interaction between two factors in a factorial ANOVA.

Fisher's exact test A statistic for 2×2 contingency tables used when the total number of observations is small.

Fisher's z transformation A method for transforming the Pearson's correlation coefficient to make it more amenable to analysis.

fixed effect This is often applied to describe a factor in ANOVA that is set by the experimenter (in contrast to a random effect).

four-way… An experiment involving four independent factors. Read entry for two-way… and extrapolate.

Friedman test A non-parametric test that is a restrictive version of a two-way ANOVA. Only one observation is allowed for each factor combination.

frequency The number of times a value (or range of values) occurs in a sample; for example, there were 56 white-eyed flies in the vial.

frequency distribution The number of times each possible value occurs in a sample.

function Some (hidden) R code that carries out a test or transformation etc.

_G_-test A form of contingency table test; a.k.a. log-likelihood ratio test.

gamma distribution A family of continuous distributions.

general linear model (GLM) A term used to describe a family of analyses that include ANOVA and linear regression.

generalized linear model A synonym for general linear model.

Genstat A statistical package (capable of carrying out complicated ANOVA designs).

geometric mean A measure of position: the antilog of the mean of the logs of each value.

geometric mean regression A model II regression method.

GLIM A statistical package very capable in general linear models.

goodness of fit Any statistic that compares the actual distribution of a variable with a theoretical one.

graphical representation Any method of displaying data visually.

group A set of observations with something in common; for example, all from the same tree.

grouping variable A variable by which observations can be divided into groups; for example, tree number.

harmonic mean A measure of position: the reciprocal of the mean of the reciprocals of each value.

heteroscedasticity When different groups have unequal variance and therefore the data violates the assumptions of ANOVA and regression.

hierarchical ANOVA A synonym of nested ANOVA.

highly significant Where the null hypothesis is rejected because the P-value is much less than the 0.05 level. The actual level varies with authors but P must be less than 0.01.

histogram A graph to display the distribution of a continuous variable where each category on the x-axis represents a range of values.

homogeneity of variances An assumption of many parametric tests such as the t-test or ANOVA is that the variances in the groups are equal. This can be tested using the Levene test.

homoscedasticity When different groups have equal variance and therefore conform to the one of the assumptions of ANOVA and regression.

Hotelling's trace A method for calculating P-values in MANOVA.

hypothesis This is what is being tested when a statistical test is carried out; for example, 'male and female tree frogs have a different mean weight'.

independence An assumption of many statistical tests. The data collected in a sample are not affected by whether another event has occurred. Many statistical tests are rendered invalid because of lack of independence.

independent samples No individuals measured in sample A are also measured in sample B. If this is violated, the sin is called pseudoreplication.

independent samples *t*-test *t*-test comparing two sets of data that are not paired.

independent variable The 'cause' variable set or varied by the experimenter.

individual In statistics an 'individual' can be a pair, bone, region, species or any number of different things. The 'individual' will provide one observation for each variable under consideration and in when the data are arranged in the package will provide a single row.

integer A whole number.

interaction In ANOVA this is a measure of whether two or more grouping variables have an additive (no interaction) effect or not (interaction).

intercept In regression analysis; the point on the y-axis when the value of x is zero.

interquartile range A measure of spread: if the data are in rank order it uses the range from the value 25% down the list to that 75% down.

interval data When observations are made on a meaningful measurement scale.

jack-knifing A technique to determine bias in statistics by recalculating using a subset of the data.

Kendall rank correlation A non-parametric test to measure correlation (association).

Kendall robust line fit A non-parametric, model II, version of regression (rarely used).

Kolmogorov–Smirnov test A non-parametric test that compares two distributions. Very useful for goodness-of-fit tests and more powerful and convenient than a chi-square goodness of fit when samples are large.

Kruskal–Wallis test A non-parametric version of a one-way ANOVA that tests the null hypothesis that two or more groups come populations with the same median.

kurtosis A measure of the shape of a distribution (sometimes called g_2).

latin square An experimental set up that reduces possible bias caused by unquantified variation in the environment across an experiment.

Lawley–Hotelling test A method for calculating *P*-values in MANOVA.

least significant difference test The simplest and probably the most widely used *post hoc* test used after a one-way ANOVA to determine which groups differ from which.

legend A small section of text that should accompany *every* figure or table in a report, making it interpretable without recourse to the main text (synonym for caption).

leptokurtic A distribution that has a lot of observations around the mean and in the tails but fewer in the 'shoulders'.

level A particular treatment; a defined condition for the independent variable (set by the experimenter).

Levene test A test for homogeneity of variances. Used for checking data to see if a parametric test such as ANOVA is appropriate.

liberal test A statistical test where the null hypothesis is rejected more often than it should be.

line chart (graph) A graphical representation of data where points are joined. This makes the assumption that values between points are possible and likely to be fairly represented by the position of the line.

linear regression Regression that assumes the 'cause-and-effect' relationship is best described by a straight line.

log-likelihood test A method of calculating the probability that a particular contingency table will arise; it is the basis of the G-test.

log transformation An operation where each observation is logged (usually to base 10).

logistic curve A distribution, usually with a range from 0 to 1, where the values near the extremes are approached with asymptotes.

logistic regression A version of regression where the 'effect' variable is transformed using logits. Useful when using regression to predict values that have a restricted range of possible values (e.g. proportions or percentages).

logit transformation Used to convert values on a scale with limits (e.g. proportions have limits at 0 and 1) to a limitless one. Transformation for a proportion, x: Logit $x = \ln(x/(1-x))$.

LSD test Least significant difference test; a commonly used *post hoc* test, used after a one-way ANOVA to determine which groups are different from which.

main effect In ANOVA a factor (grouping variable) that is at the top of a design (i.e. not nested).

MANCOVA A MANOVA that includes a covariate.

Mann–Whitney U test A non-parametric test of a null hypothesis that two groups come from the same distribution. A synonym of Wilcoxon–Mann–Whitney test and Wilcoxon signed ranks test.

MANOVA An acronym for multivariate analysis of variance. A test where there is more than one dependent variable under investigation.

Mantel test A matrix method for determining how two matrices of data are associated.

matched samples/data A synonym for paired samples or repeated measures.

mean A measure of position: sum of all values divided by the number of observations (synonym of arithmetic mean and average).

mean square (MS) Used in ANOVA; the sum of squares divided by the degrees of freedom (SS/df).

measured variable A variable where the experimenter has to take a reading; for example, height, weight, optical transmission.

measurement An observation or single item of data (datum).

meta-analysis A method for combining the results of several tests of the same hypothesis. It can be used even when the original data are not available and different tests have been used.

median A measure of position: if all the data are put in rank order it is the value of the datum in the middle.

MINITAB A statistical package.

mixed model (ANOVA) A test with both fixed and random grouping variables (factors).

mode A crude measure of position: the most commonly occurring value.

Model I ANOVA The basic ANOVA where the grouping variables (factors) are all fixed effects.

Model II ANOVA An ANOVA where all the grouping variables (factors) are random effects.

Model I regression The usual parametric regressions that assumes the 'cause' variable is measured without error.

Model II regression A rarely used version of regression that takes into account the fact that the 'cause' variable may be measured with error.

multifactorial design There are many variables that can be used to group the data.

multiple correlation Difficult to interpret comparison of more than two variables.

multiple regression A test that establishes the best prediction of an 'effect' variable using all 'cause' variables simultaneously.

multivariate statistics Tests which use more than one dependent variable.

negative binomial distribution A discrete probability distribution which is frequently invoked to describe contagious (clumped) distributions.

nested ANOVA Synonym of hierarchical ANOVA. A test where at least one of the grouping variables is a subgroup of another; for example, 'bunch' as a grouping variable within the grouping variable 'vine' in an experiment on grapes.

nominal Where the values in a data set cannot be put into any meaningful sequence only assigned to categories (e.g. blue and red).

non-parametric test A test where few or no assumptions about the shape of a distribution are made.

normal distribution A unimodal, continuous probability distribution with a characteristic bell shaped curve. This distribution is often assumed of the data in parametric statistics.

null hypothesis Every hypothesis being tested must have a null hypothesis; for example, if the hypothesis is that two groups have different mean heights then the null hypothesis must be that the two groups do not have different mean heights.

observation A single item of data (datum); a measurement.

observer bias Whenever two, or more, observers have differences in the values recorded from the same observations. Can often be a consistent difference that can be corrected for.

one-tailed test A test that assumes rejection of the null hypothesis can only come from a deviation in one direction rather than either. For example, the hypothesis that two groups are different is 'two-tailed' while the hypothesis that group 'A' is larger than group 'B' is 'one-tailed'. The effect of using a one-tailed test is to make statistics much less conservative for the same value of P.

one-way ANOVA A parametric test of the null hypothesis that two or more groups come from the same population.

ordinal When values in a set of data can be placed in a meaningful order and ranked.

orthogonal Literally means 'at right angles to'. In statistics it is used to indicate that two variables, factors or components are unrelated to each other.

outlier An extreme or aberrant observation lying well away from the rest of the data.

P-value The probability of the significance statistic being that extreme or more if the null hypothesis is true. In biology the null hypothesis is usually rejected if the P-value is <0.05. Often written as p-value.

paired observations Two observations taken from the same individual, perhaps in a 'before and after' design (synonym for repeated measures).

paired samples Sets of data comprising paired observations.

paired t-test To test the null hypothesis that two set of observations on the same individuals (e.g. before and after) have the same distribution.

parametric test One where assumptions about the shape and spread of the data are made within the test.

partial correlation A method of examining relationships between more than two variables that examines them pairwise while other variables are held constant.

path analysis A complex data-exploration system where there are several simultaneous and possibly interacting 'cause' and 'effect' relationships.

Pearson product-moment correlation coefficient The standard parametric correlation coefficient, r, measuring the association between two variables, that varies from 1 (perfect positive correlation), through 0 (no relationship) to −1 (perfect negative correlation).

percentage When the relationship between two values is expressed as a single value (usually on a scale from 0 to 100).

phi coefficient The result of a 2×2 contingency table.

pie chart A simple representation of frequencies of observations in different categories as sections (slices) of a circle. Particularly appropriate when the categories do not have a logical sequence.

Pillai's test/Pillai's trace A method for calculating P-values in MANOVA.

platykurtic A distribution that has a more observations in the 'shoulders' and fewer around the mean and in the tails than a normal distribution.

Poisson distribution A discrete frequency distribution that results when events occur entirely at random.

polynomial regression A regression where the relationship between 'cause' and 'effect' is not assumed to be a straight line.

population The pool of possible individuals from which a sample is taken. Do not confuse with a biological population which will include additional individuals.

position The position of the sample is its mid-point, which can be defined as a mean, median or mode.

***post hoc* test** Meaning a test 'after this'; there are several tests used after a one-way ANOVA to determine which groups are different from which. Common methods include LSD, SNK and Bonferroni.

power analysis A method based on the number of observations for determining the likelihood of detecting a statistically significant effect.

precision The range of possible values between which a particular observation may lie. The repeatability of a measurement (different to accuracy which is a measure of the closeness of an measurement to the true value).

prediction interval In regression the range of y values that are expected for a given value of x. Usually given as a 95% prediction interval. This range does not fall within a zone running parallel to the best-fit line in linear regression but is smaller around the mid-range of x.

predictor In regression the 'cause' variable is often called the predictor and is always plotted on the x-axis.

principal axes A synonym for principal components.

principal axis regression A model II regression technique.

principal component analysis A multivariate test which weights the variables to maximize the differences between individuals.

probit A transformation of data based on the use of cumulative probabilities and log graph paper.

procedural control A control where all the disturbance associated with applying a treatment is carried out, but the treatment itself is not applied.

proportion When the relationship between two values is expressed on a scale from 0 to 1 (e.g. 30 out of 60 becomes 0.5).

proportional frequency A synonym for relative frequency.

pseudoreplication A problem in statistics when samples that are not independent are being treated as such.

q-q plot A q-q or quantile-quantile plot is a graphical method for assessing whether a variable follows a normal distribution.

quadrat A square sampling area.

quadratic regression A regression where the relationship between the variables is assumed to be best described by a quadratic equation.

qualitative A observation that is assigned to a category that, although it may be coded as a number, has no numerical value (e.g. sex: coded 1 for female and 2 for male).

quantitative An observation that has a meaningful numerical value. It can be either a direct observation or a count.

quartile When the data are ranked the quartiles are the values of the data points 25% and 75% down the list. They form the limits for the interquartile range.

r The result of a Pearson product-moment correlation. If $r=0$ there is no correlation. If $r>0$ there is a positive relationship and if $r<0$ it is negative, 1 is perfect correlation, -1 is perfect negative correlation.

r_s The statistic associated with the Spearman's rank correlation test.

R A free version of the statistical package S.

random effect A term applied to factors in ANOVA that are not set by the experimenter (in contrast to a fixed effect).

random sample A sample where each individual in the population has an equal chance of being measured or collected.

randomized block design When sampling units are placed into groups (blocks) and the treatment applied to each sample is randomized within the block.

range A crude measure of dispersion: the distance from the lowest to highest value in a data set.

ratio When two values are expressed as a single number (e.g. 6:2 becomes 3). Ratios lose information and magnify the error associated with measurement.

raw data Observations as they were originally recorded before any transformations or other processing is applied.

reduced major-axis regression A model II regression technique where the slope is essentially determined by the ratio of the standard deviations of the x and y values.

regression A description of the relationship between two variables where the value of one is determined by the value of the other (synonym of linear regression). More advanced regression can use several 'cause' variables.

related measures A synonym for paired samples or repeated measures.

related samples A synonym for paired samples or repeated measures.

relative frequency The proportion of observations having a particular value (or range of values). It is the frequency scaled to the sample size (e.g. 45 of 108 nests in a survey had four eggs, the relative frequency of four eggs is 45/108 or 0.417 or 41.7%).

repeated measures Two or more observations taken from the same individual, same site, same transect, etc., at different times (if only two observations then this is a synonym for paired samples).

repeated-measures ANOVA ANOVA carried out using repeat observations of the same individual. Time of observation (e.g. before and after) will be used as one of the factors in the ANOVA but the degrees of freedom will be reduced.

residuals The variation in the data left over after a statistical model has been accounted for (often regression or ANOVA). The model with the best fit has the smallest residual variation.

response In regression the 'effect' variable is often called this and is always plotted on the y-axis.

Ryan–Joiner test A method for determining whether a set of data follows a normal distribution.

S A statistical package (also comes in S-plus version that has a graphical user interface).

sample As all the individuals in a population may rarely be counted a portion of the population has to be taken, this is a sample.

sample size The number of observations in a sample.

sample variance The variance of a single sample.

sampling unit The level at which an individual observation is made; for example, a quadrat or a given size; a single leaf.

SAS A widely used and powerful statistical package.

scatter (plot) A graphical method for examining two (or possibly three) sets of data for possible relationships.

Scheffé-Box test A test for heterogeneity of variance.

Scheffé test One of many *post hoc* tests used in ANOVA to determine which groups are different from which.

Scheirer–Ray–Hare test A weak, non-parametric analogue of a two-way ANOVA, rarely supported in packages but quite easy to implement using the usual parametric ANOVA on ranked data and simple treatment of the resulting F-value.

second-order interaction An interaction between three factors in ANOVA.

Sidák test A synonym for the Dunn–Sidák test for multiple comparisons.

sign test A very conservative non-parametric test of a null hypothesis that there is no difference between two groups.

significance level The probability of achieving a significant result if the null hypothesis is true. In biology this is usually set at 0.05.

significant When the null hypothesis is rejected because the P-value is less than 0.05 (the usual value in biology), 0.01, or any value set by the tester.

simple factorial (design) None of the grouping variables are subgroups of any others; i.e. there is no nesting.

single-classification ANOVA A synonym for one-way ANOVA (there is only one grouping variable).

skew(ness) A measure of the symmetry of a data set (sometimes called g_1). Positive skew indicates that there are more values in the right tail of a distribution than would be expected in a normal distribution. Negative skew indicates more values in the left tail.

skewed distribution A distribution that has a value of skewness other than zero.

slope A number (usually b or β) denoting how a trend line deviates from zero (a slope of 0).

SNK test Student–Newman–Keuls test, a common *post hoc* test.

Spearman rank-order correlation A non-parametric measure of correlation.

Spearman's rank correlation An alternative name for Spearman rank-order correlation.

split-plot design An experimental design technique used to analyse two factors when there is only one 'plot' for each level of one of the factors.

spread The way in which the data are distributed, often measured by standard deviation (synonym of dispersion).

SPSS A widely used statistical package.

standard deviation A measure of spread: sensitive to shape of distribution.

standard error A measure of spread: the standard deviation of the values of a set of means taken from a data set. Sensitive to sample size.

Statistica A statistical package.

stem and leaf chart A method of displaying the data commonly used by computers before graphical output was possible. The 'stem' would be represented by a series of rows and leaf by columns starting from the left. Each observation would be assigned to a position on the stem based on its value.

stepwise regression A regression analysis where the best method for predicting the 'effect' from several 'cause' variables is sought.

stratified random sample A method of collecting a sample that takes into account a feature of the collecting area.

Student A pseudonym used by the statistician William Gossett.

Student's *t*-test A synonym for independent samples *t*-test.

Student–Newman–Keuls (SNK) test A frequently used *post hoc* test, used after a one-way ANOVA to determine which groups are different from which.

summary statistic Anything that condenses the information about a variable, such as a mean or standard deviation.

symmetry A data set with symmetry has the same shape either side of the mean.

Systat A statistical package.

t-distribution A family of distributions widely used in statistics that is derived from the distribution of sample means with respect to the true mean of a population.

t-test To test the null hypothesis that two groups come from the same distribution (synonym for Student's *t*-test, independent samples *t*-test).

tally When observations are assigned to categories and marked as ticks in a table.

three-way... An experiment involving three independent factors. Read entry for two-way... and extrapolate.

time series A set of data points taken at different points in time.

transect A method of taking a sample of observations. Usually by selecting a straight line between two random points, or from one random point in a random direction for a set distance.

transformation A mathematical conversion that is applied to every observation in a data set. Usually used to make a distribution conform to a normal distribution.

treatment A level (usually denoted by an integer) of an independent variable, factor or grouping variable (i.e. set or defined by the experimenter).

Tukey test One of many *post hoc* tests used in ANOVA to determine which groups are different from which.

Tukey–Kramer method A synonym for Tukey test. A *post hoc* test, used after a one-way ANOVA to determine which groups are different from which.

two-tailed test Applies to most statistical tests and implies that the null hypothesis can be rejected by deviations either up or down. For example, if the null hypothesis that two groups of bats use the same frequency for echo location is rejected then group 'A' may use either a significantly higher or significantly lower frequency than group 'B'. If the standard $P=0.05$ level is used then it implies a $P=0.025$ region in each tail.

two-way ANOVA An ANOVA test where there are two independent ways of grouping the data (two factors).

two-way interaction In ANOVA this a measure of whether two grouping variables have an additive (no interaction) effect or not (interaction).

type I error When a truly non-significant result is deemed significant by a test.

type II error When a truly significant result is deemed non-significant by a test.

unbalanced When there are different numbers of observations in different factor combinations. Severely unbalanced designs (i.e. where some of the factor combinations have no observations) should be avoided.

uniform distribution A 'flat' distribution where the chance of any value occurring is approximately equal, may often be transformed, using the arcsine transformation, to an approximately normal distribution.

unimodal A frequency distribution with a single peak at the mode.

univariate statistics Statistical tests using only one dependent variable.

unpaired data A synonym for independent data, stressing that sets of data are not paired.

value A single piece of data (datum).

variable Anything that varies between individuals (e.g. 'sex', 'weight' or 'aggressiveness'). The term variate is actually correct, but variable is now the widely used term for the observed data set.

variance The sum of squared deviations of observations from the mean: a measure of spread of the data. Very important in the mechanics of statistics but not very useful as a descriptive statistic.

variance/mean ratio (v/m or s^2/m) A commonly quoted descriptive statistic useful for determining whether a set of observations fits a Poisson distribution ($v/m=1$), is more clumped ($v/m>1$) or is more ordered ($v/m<1$).

variate The correct term for variable. Still retained for terms such as canonical variate analysis or univariate statistics.

Weibull distribution A family of continuous distributions.

Welch's approximate *t*-test A version of the Student's *t*-test that can be used when the variances of the two samples are known to be unequal.

Welsch step-up procedure A *post hoc* test, used after a one-way ANOVA to determine which groups are different from which; requires equal sample sizes.

Wilcoxon–Mann–Whitney test, Wilcoxon rank sum W test Synonyms for the Mann–Whitney U test. A non-parametric test of a null hypothesis that two groups come from the same distribution.

Wilcoxon signed rank test A non-parametric test of a null hypothesis that there is no difference between two related groups. The non-parametric equivalent of the paired t-test.

Wilks' test A method for calculating P-values in MANOVA.

Williams' correction A method of correcting bias in various contingency tests such as the G-test.

winsorize A method used to reduce the effect of outlying observations by replacing them with the next value towards the median.

within Sometimes used as shorthand for 'within-sample variance' or 'within-group variance'.

within-sample variance In ANOVA a measurement of the amount of variation within a sample; *see* between-sample variance.

x-axis The horizontal axis of a graph or chart (abscissa).

Yates' correction Sometimes called the 'continuity correction'. A method to make the results of a 2×2 chi-square test more conservative.

y-axis The vertical axis of a graph or chart.

z-axis The axis that goes 'into' the paper or computer screen on a three-dimensional graph or chart.

z-distribution Occasionally used as a synonym for the normal distribution.

z-test A test used to compare two distributions, or more usually to compare a sample with a larger population where the mean and standard deviation are known.

Assumptions of the tests

Most statistical tests make assumptions about the data to which they are being applied. If the assumptions are violated it is wise to treat the results with caution, especially when P-values fall in the range 0.01 to 0.1.

Here is a test-by-test summary of the assumptions.

Test	Assumptions
G-test	Observations can be assigned to groups or categories
chi-square test	Observations can be assigned to groups or categories
Kolmogorov–Smirnov	Observations come from a fairly continuous scale
paired t-test	Both sets of data are normally distributed and variance is the same in both samples (although tests are often incorporated into statistical packages that make corrections by adjusting the degrees of freedom)
Wilcoxon signed ranks test	Observations are made on a scale such that the magnitude of differences is meaningful
sign test	Observations are made on a scale so that the question 'is A bigger than B?' can be answered
t-test	Both sets of data are normally distributed and variance is the same in both samples (although there are test often incorporated into statistical packages that make corrections)
Mann–Whitney U test	Observations are made on a continuous scale (i.e. they can be put into rank order with very few ties)
Friedman test	One observation per factor combination observations may be put in meaningful rank order
all ANOVA (analysis of variance) tests	Observations are independent both within and between samples
	Variance is the same in all samples
	Data are normally distributed within each factor (or factor combination)
	Observations are assigned to groups (coded by integers) using one or more factors

Choosing and Using Statistics: A Biologist's Guide, 3rd Edition. By Calvin Dytham.
Published 2011 by Blackwell Publishing Ltd.

Kruskal–Wallis test	Observations are made on a fairly continuous scale (i.e. they can be put into rank order with very few ties)
Scheirer–Ray–Hare test	Observations are made on a continuous scale (i.e. they can be put into rank order with very few ties)
chi-square test of association	Observations can be assigned to categories or groups using one or more factors
phi coefficient of association	Observations can be assigned to two groups for each of two factors
Cramér coefficient of association	Observations can be assigned to categories or groups using two factors
'standard' correlation (Pearson product-moment correlation)	Individuals have observations for two variables measured on a continuous scale Two variables are both normally distributed
Spearman's rank-order correlation	Individuals have observations for two variables measured on an approximately continuous scale
Kendall rank-order correlation	Individuals have observations for two variables measured on an approximately continuous scale
Kendall robust line-fit method	'Effect' measured on an approximately continuous scale 'cause' on any meaningful scale
ANCOVA (analysis of covariance)	Observations and covariate measured on a continuous scale Variance the same for all factor levels Residuals are normally distributed Observations are independent
'standard' regression (model I linear regression)	'Cause' (=independent or x) variable is measured without error Variation in 'effect' (=dependent or y) is the same for all values of 'cause' Relationship between x and y is linear 'Effect' is measured on a continuous scale 'Effect' should be normally distributed for any value of 'cause'
logistic regression	'Cause(s)' (=independent or x) variable(s) measured without error, can be categorical variable(s) Variation in 'effect' (=dependent or y) the same for all values of 'cause' Relationship between x and y is linear 'Effect' can be expressed as a proportion (and then transformed by logits), can be a categorical variable
model II regression	Individuals have observations for y variable measured on an approximately continuous scale
polynomial regression	As standard regression but not assuming that the relationship between x and y is linear

stepwise regression/ multiple regression/ path analysis	As standard regression but with several 'effect' variables measured for each individual
discriminant function analysis	Individuals have two or more observations assigned to them measured on continuous scales
principal component analysis or factor analysis	Individuals have two or more observations assigned to them measured on continuous scales
canonical variate analysis	Individuals have two or more observations assigned to them Observations are measured on continuous scales Individuals can be assigned to groups
MANOVA (multivariate analysis of variance)	Two or more observations for each individual Observations are independent both within and between samples Observations are assigned to groups (coded by integers) using one or more factors Variance is the same in all samples Residuals are normally distributed
MANCOVA (multivariate analysis of covariance)	Two or more observations for each individual Observations are independent both within and between samples Variance is the same in all samples Residuals are normally distributed Observations are assigned to groups (coded by integers) using one or more factors Covariate is measured on a continuous scale
cluster analysis (a family of techniques that have slightly different assumptions)	Each individual has two or more observations assigned to it Observations are measured on meaningful scales Individuals can be assigned to groups

What if the assumptions are violated?

There are several possible courses of action that can be taken (in approximate order of preference):
1. data could be transformed to make them suitable for the analysis chosen;
2. an alternative test of the same hypothesis but with different assumptions is used instead;
3. the hypothesis is reframed to allow a different test to be used;
4. violation of the assumptions could be ignored totally but the results regarded with caution;
5. no test is carried out at all.

Hints and tips

Using a computer

- Save frequently: computers crash and storage media of all kinds fail every now and again and you want to make sure you don't lose data.
- Learn a few keyboard shortcuts.
- An easy way to select a block of text or data in many packages is to place the cursor at the beginning, move the pointer to the end and press Shift as you left-click the mouse.
- Another way to select blocks of text is to hold Shift while moving the down arrow, up arrow, Page Up or Page Down.
- Using the underlines: the underlined letters in menus mean that you can access the menu by typing the letter on the keyboard while holding the Alt key.
- Use the Tab key to move between boxes: useful in many of the Windows dialogue boxes.
- Use Shift and Tab together to move backwards through boxes: useful to correct mistakes.
- Back-up your important files frequently on memory stick, CD, web storage, etc., and keep physical back-ups in a different place to avoid total loss from theft or fire.
- Holding Alt and pressing Tab moves you between open packages.
- Edit in the best editing package, then do the statistics or graph drawing in another: do not feel that you have to use the pathetic spreadsheet capabilities of the statistics package.
- If you are given data in the format of another package that your package cannot read you can nearly always read it by saving in raw text format from the first package.
- When converting labels into numbers, using alphabetical order all the time will avoid many problems of converting the numbers back to labels.
- Cut and paste is a very powerful facility of most packages: you can usually copy material from one to another using copy and paste.
- The keyboard shortcuts for cut, copy and paste are nearly always Ctrl+x, Ctrl+c and Ctrl+v respectively. Using the shortcuts is easier and quicker than going to the Edit menu and selecting from there.

Choosing and Using Statistics: A Biologist's Guide, 3rd Edition. By Calvin Dytham. Published 2011 by Blackwell Publishing Ltd.

- Double-clicking or right-clicking often brings up helpful options.
- In Excel, clicking the plain square on the top left of the spreadsheet between A and 1 selects all cells and allows you to change all fonts or column widths, etc.
- If you get stuck try the help file: these are usually extensive and often have examples too.
- Alt + F4 will usually close a package.
- Don't leave lots of unnecessary windows open: they slow the computer down and clutter the environment.

Sampling

- Try to balance sampling designs if you can (i.e. take equal-sized samples).
- Measure everything you can easily: you never know what is going to be important.
- Avoid sampling at regular intervals if possible.
- Choosing the nearest individual to a random point will *always* bias the sample to individuals on the edge of clumps and against those in the middle.
- Don't carry out repeat sampling in the same sequence.
- Measure to sensible precision only, not to maximum, but make sure there are at least 30 different possible values wherever possible.
- Check the quality of measurements by repeat measuring the same individual after an interval.
- Three subsamples from a site are much better than two.
- When setting up a laboratory experiment make sure it is genuinely factorial (i.e. there should be no confounding factors such as 'all species *x* was from the sunny site').
- Randomize measurement routines as much as possible (i.e. don't measure all of group 1 then 2, etc.).
- Try double-blind labelling if possible (i.e. when measuring you don't know what group the individual belongs to).
- Don't design over-elaborate experiments: it is difficult to interpret anything with more than three factors.
- Use transects with caution as they can easily produce biased samples.
- If measurements are taken by several different people check the quality of the data by having everyone blind measure the same individuals.
- Always sample with a clear idea of the statistical test you intend to use in mind.

Statistics

- Try the analysis on dummy data before collecting any real observations.
- Find a worked example of a design like the one you are doing in a statistics book and try to repeat the result in your statistical package.
- Frame null hypotheses very carefully before anything else.

• Always consider whether the data violate the assumptions of the test: if they do, be wary of the results.
• Transformation of the data can often turn an inappropriate data set into an appropriate one.
• One-tailed tests have their place (i.e. the alternative hypothesis is 'x is greater than y' rather than 'x is different to y') but if in any doubt use two-tailed tests.
• If P-values are close to 0.05 consider resampling to increase sample sizes.
• There is nothing 'special' about $P = 0.05$, so don't be completely tied to it.
• A P-value of 0.05 means a one-in-20 chance of getting a result this, or more significant, even if the null hypothesis is true.
• In regression, if you are unsure which variable is the 'cause' and which is the 'effect' then the data are probably not suitable for regression anyway.
• If a non-parametric test with reasonable power is available use it.
• Carry out tests on incomplete data sets to get a feel for the results from the complete set.
• Use power analysis to help inform you as to the potential effect of further sampling.
• Use 95% confidence intervals rather than standard errors when comparing several means.
• The coefficient of variation is a good way to compare data sets with very different means.

Displaying the data

• Never use three-dimensional effects for bar charts, pie charts, etc. (except, possibly, for posters).
• If you must use a three-axis graph make sure that every point is anchored to the 'floor' by a spike, otherwise there is no way of determining its position on *two* of the axes.
• Use the minimum amount of shading.
• Use black and white rather than colours (except, possibly, for posters).
• Avoid putting titles on graphs and figures.
• Use a figure legend for every graph and make sure that the legend is informative enough to make the graph intelligible without reading the main text of a report.
• Use a different font, font size or margins to differentiate figure legends from the main text.
• Make sure figures and tables are appropriately numbered and referenced correctly from the text.
• Don't use any more decimal places than you have to and, for raw data, no more than you have measured.
• If a graph has a measure of position (e.g. mean) then nearly always display a measure of dispersion as well (e.g. standard deviation or 95% confidence interval); if plotting medians then always plot quartiles too.

- If you want the reader to compare figures make sure they have the same scales if possible.
- If you use a line graph it must be possible for intermediate values to exist as they are implied by the line.
- Don't be afraid to use log scales even when the observations are not logged, and remember that $\log_{10}$ is easier for a reader to mentally convert back to the original value than natural log.
- Never draw best-fit lines unless the data are suitable for regression.
- Never extend best-fit lines beyond the range of the data.
- Always have gaps between bars on a bar chart (data are discrete).
- Never have gaps between bars in a histogram (data are continuous).
- Only use pie charts for categorical data that have no real scale.
- Don't clutter graphs with too much information.

A table of statistical tests

Choosing and Using Statistics: A Biologist's Guide, 3rd Edition. By Calvin Dytham.
Published 2011 by Blackwell Publishing Ltd.

Tests of difference

Samples or groups	Factors	Data type	Fit to known distributions	One-sample tests
1	–	Cat	G-test	fit to uniform: G-test, chi-square test
		D	fit to Poisson: chi-square test	e.g. median of 0?: Wilcoxon's one-sample test
		C	fit to normal: Kolmogorov–Smirnov test, Anderson–Darling test	e.g. mean of 0?: one-sample t-test
			Unpaired data	**Paired data**
2	1	Cat	chi-square test	chi-square test
		D	Mann–Whitney U test	Wilcoxon's signed ranks test
		C	t-test, one-way ANOVA	paired t-test
>2	1	Cat	chi-square test	chi-square test
		D	Kruskal–Wallis test	Friedman test for repeated measures (no replication)
		C	one-way ANOVA	repeated-measures ANOVA
2+	>1	D	(2 factors only) Friedman test (if no replication), Scheirer–Ray–Hare (weak option)	Friedman test for repeated measures (only one factor other than repeat and no replication)
		C	two-way ANOVA, or multiway ANOVA	repeated-measures ANOVA

Analysis with covariate(s)

Samples or groups		Data type	
>1 group, 1+ factor, 1+ covariate	1 variable	ANCOVA	
	>1 variable	MANCOVA	

Tests of relationships

Correlation

Samples or groups	Data type	Correlation
2	Cat	chi-square for association
	D	Kendall's rank correlation, Spearman's rank correlation
	C	Pearson product-moment correlation

Regression

Samples or groups	Data type	Regression
1 cause, 1 effect	D	logistic regression, model II regression, Kendall's robust line fit
	C	linear regression, quadratic or polynomial regression
>1 effect, 1 cause	C	multiple regression, stepwise regression

Multivariate tests

Samples or groups	Data type	Multivariate tests
1+ group and any number of factors	many causes, many effects to explore	path analysis
	many variables to explore	PCA
	groups to discriminate with many variables	CVA, discriminant function analysis, MANOVA, multiple regression, DCA
	groups to discriminate with discrete variables	TWINSPAN
	many proportion or categorical variables to explore	multiple logistic regression

Note: Cat indicates categorical data; D indicates discrete or ordinal data; C indicates continuous data; although many tests will have an assumption that the data is normally distributed. CVA, cannonical variate analysis; DCA, detrended correspondence analysis; PCA, principal component analysis.

Index

Page numbers in **bold** refer to tables and those in *italic* to figures.